MW01174654

# Living with Wildlife

# Living with Wildlife

Eva Murray

Photography by
Ken Griffiths

*To my family, without whose patience and understanding this book could never have been written.*

## ACKNOWLEDGEMENTS

I wish to thank the following people, who have devoted their time and expertise towards making this book possible. From the Australian Museum, Dr Allen Greer, Walter Boles, Dr Tim Flannery and Dr Mike Gray; veterinary surgeons Teri Bellamy, Henry Hirschhorn, James H. Gill, Colin Pinney and Dr G.R. Ross; and from the National Parks and Wildlife Service of New South Wales, Greg Siepen; also the staff from Featherdale Animal Park, particularly Agata Pasqualini and Rick Webb; from Gosford Reptile Park, John Weigel and Grant Husband; the staff of St John Ambulance Australia, particularly Steven J. Johnson and Perry Neckeraurer; from the New South Wales Agriculture and Fisheries Department, Bruce White, Roger Toffolon and Grant Delves; from the Forestry Commission of New South Wales, Ted Taylor and Dick Turner; from Taronga Zoo, Les Clayton; from Flick Pest Control, Peter Lamond; from The Society for Growing Australian Plants, Judy Smith and Glen Harvey; from Wildlife Information and Rescue Service, Mikla Lewis; from Koala Park Sanctuary, David McNamara; from the Australian Herpetological Society, David Giblin; from the Koala Preservation Society of New South Wales, Jean Starr; from the Australian Pigeon Fanciers Protection Association, Nicholas Kaparos; beekeeper, N.J. Williamson; herpetologists, Ken Griffiths and Paul Levy; from AWARE, Karen Fenn; from the Sydney County Council, John Findlayson; Dr Douglas James Bone; naturalist, John Cann; racing-pigeon fancier, Jack Cope; from Qantas, Ken Lewis; from the Australian Veterinary Association, Colonel Warren Bassam; from the Department of Health New South Wales, Dr P. Christopher; Entomologist Phillip Hadlington.

First published 1989 by
**Reed Books Pty Ltd**
**2 Aquatic Drive Frenchs Forest NSW 2086**



Edited by Michelle Wright
Designed and illustrated by John Windus
Typeset in New Zealand by Saba Graphics Ltd.
Printed in Singapore for
Imago Productions (FE) Pte Ltd

National Library of Australia
Cataloguing-in-Publication data

Murray, Eva, 1952-
Living with Wildlife

Includes index.
ISBN 0 7301 0286 6

1. Urban fauna – Australia.
2. Wildlife rescue – Australia. I. Title.

591.52'68'0994

# CONTENTS



**DID YOU KNOW?**

Many animals, including reptiles, are protected by law and it is illegal to kill them unless they are a threat to life, stock or property. Prior to handling or keeping any protected native animals, a licence must be obtained from your State Fauna Authority. There is no charge and arrangements can be made by presenting yourself to the department, in writing or by phone. If help has been given to an animal in an emergency and you are nursing it obtain the licence as soon as possible. It is legal to keep some species of animals without a licence. Contact the department to find out for which species you need a licence in your State.

# FOREWORD

How do we humans live harmoniously with wildlife? Sometimes it's extremely difficult, especially if the animal in question is gnawing through the television cable at a crucial time during a serial. On the other hand, the idea of birds, lizards, and even possums in the garden is attractive, particularly to urban dwellers.

*Living with Wildlife* will tell you how to encourage wildlife to your garden, as well as how to cope with problem animals, by understanding the animals' habits, where they normally live and their peculiarities.

Eva Murray does this in an easy-to-read style. She has experienced most of the traumas associated with marauding wildlife, and it was her experiences with peach-stealing fruit-bats and wasps invading the kitchen that prompted her to write this book for all people who come in close contact with wildlife.

Poisonous animals such as spiders and snakes are covered thoroughly, especially first-aid procedures and ways of avoidance. You will also find advice on how to help sick or injured animals, and where to go for advice and assistance.

In compiling this book Eva consulted numerous experts, including the Australian Museum, animal parks, the NSW National Parks and Wildlife Service and other state wildlife authorities, so as to present the most up-to-date information. It is the perfect book to have around the house.

Ian Sim
Acting Director
National Parks and Wildlife Service

# INTRODUCTION

Everyone, at one time or another, comes in contact with wildlife. Although such encounters are common in areas rich in vegetation, they may be more disturbing in areas where wildlife is not expected or understood.

This book has been written to serve as a guide for anyone who encounters the unexpected from wildlife. It is written by a layman for the laymen of nature, using simple terms, and Latin names only as part of the identification of the species.

The aim of the book is to encourage a harmonious co-existence between people and wildlife. That may entail knowing what to do to help injured wildlife, or knowing how to avoid being injured by wildlife. Lovers of wildlife, and people inconvenienced by it, should find this book equally useful. Though I have tried to remain unbiased, I remain against the senseless killing of wildlife that is often brought about by the fear of the unknown. I would also not encourage the keeping of native animals except when they need temporary care due to injury, sickness or being orphaned.

Throughout my research, one point emerged quite clearly. When it comes to wildlife, opinions vary, particularly on the subject of feeding and rearing animals. Procedures suggested in this book are a guide only, and different opinions should be taken into consideration. Use your judgement, keeping in mind that if it is not damaging to the animal and it works, you have succeeded. If you tried, you have not failed.

Let me present this book to all of you, who every day wake up to the face of nature and its grand adventures into the unknown.

Eva Murray

# HOW TO USE THIS BOOK

This book deals with the most common wildlife encountered in urban Australia. It is not possible to include every species. Commonsense should be used when encountering species not covered in the book. Find a species most closely resembling the one you are faced with, not so much in looks but in its habits in the wild. Use the main heading as your first guide. For example, procedures under 'Ducks' can be applied to all ducks, not just the Grey Teal, which has been singled out and written up in greater detail. Particular species of wildlife, for example Grey Kangaroos, are often further divided into particular varieties. Such varieties are usually similar in habits and for the purpose of this book, can be treated as one.

Further identifiers, such as whether the animal is native, nocturnal, territorial, or migratory, should help to anticipate when the animal is most likely to be encountered, when it is most likely to accept food, when is the best time to release it, whether the animal has the preferential status of a native, or whether it is an introduced species and a possible pest, and if the animal needs to be housed in a pouch.

Under IDENTIFICATION will be found a description of the size of the species in question, and any distinguishing features it may have. An identifying picture is included with each entry. The juvenile of some species is also described. This is particularly useful for birds whose juvenile appearance often differs markedly from that of the parents.

HABITAT indicates where an animal is likely to be encountered; the most favourable environment for housing the animal in captivity, and the most suitable spot for releasing it.

BREEDING will indicate when and where to expect offspring if you are already playing host to the adult of the species. It is also useful for determining whether an orphaned animal is still dependent on its parents.

Use the FOOD section as your first guide for feeding the animal in captivity. If possible, offer food listed in this section or food closely resembling types eaten in the wild.

EMERGENCY FOOD lists are intended as a guide only; foods suggested by other people experienced in animal rearing should also be considered. For short-term feeding, getting the animal to accept food is of primary importance, the type of food offered being secondary. For long-term feeding, further research may be needed to ensure the animal is getting proper nourishment, with necessary protein and energy supply, for its recovery.

HOSTILE BEHAVIOUR is included under species which are dangerous or feared. It is to be used to identify the animal's intention and to avoid a potentially dangerous situation.

HOW TO ATTRACT is a section for people who wish to show that extra kindness to wildlife. It lists ways to encourage the species to grace the yard with its presence. Attracting wildlife to a yard, however, also depends on factors over which the reader may have no control. For example, the species must be found naturally in the area; the attitude of neighbours is also important—they may be discouraging wildlife, while you are trying to encourage it. Unfortunately, some areas do not lend themselves to wildlife, no matter what incentives are offered.

HANDLING is intended as a guide for preventing injury to the animal—and the handler.

TRANSPORT lists suggestions for means of transporting the animal in an emergency, such as a trip to a vet. It offers solutions for housing the animal on a temporary basis. Further enquiries should be made for long-term care.

DID YOU KNOW? will acquaint the reader with the unexpected about the animal—its oddities, and its good and bad points.

In addition to the individual animal entries there are three other sections to the book.

At the front FIRST AID offers advice and suggests treatments for any injuries likely to be inflicted by animals listed in the book. As prevention is better than cure, a tetanus needle is recommended for humans after all injuries resulting from animal bites and scratches. It is also advisable to have a tetanus injection when confronting animals on a regular basis.

At the beginning of each chapter there is a section which provides solutions to common problems encountered with animals in that chapter — problems created by man, such as injuries to animals caught in electrical wires; problems created by animals, such as injuries to animals by cats; problems created by nature, such as birds falling out of the nest, and problems created for man by nature, such as animals destroying property.

At the back of the book EXTRA HELP suggests places where more information can be obtained on wildlife. It gives guides to groups, organisations and government bodies which may be able to help in solving some problems with wildlife. Further research may be needed, as it is impossible to list every animal-oriented organisation throughout Australia.

# FIRST AID

Information approved by
St. John's Ambulance Australia

## Mammals

**If bitten or clawed by an animal:**
Wash your hands. Clean the wound with antiseptic. If there is persistent bleeding, apply pressure with the fingers and elevate the wounded part. Apply a clean pad or dressing. A tetanus needle is most advisable, particularly with rats and mice, which are known to carry diseases. If the wound continues to bleed, or if it becomes pussy, swollen, or does not begin to heal after 48 hours, seek medical advice.

**If kicked by a kangaroo**:
Rest the injured part. Apply ice or cold packs to the bruised area. Apply compression bandage to minimise the swelling on limbs. Seek medical advice if concerned.

**If a kangaroo's kick winds the patient**:
Help the victim to find the most comfortable position for breathing. Do not rub or massage the victim.

**If the kangaroo's kick, or a knock with its tail results in unconsciousness**:
Position the victim on his/her side, with one arm to the side, the other across the chest. Ensure the breathing passage is clear. Loosen tight clothing. Seek medical help.

**Should the spikes or spur of an echidna penetrate someone's skin**:
Wash your hands. Clean the area with antiseptic. Should the affected area become pussy, swollen or inflamed, seek medical advice.
Note: echidnas' spurs are non-venomous, unlike that of the platypus.

## Birds

People handling birds commonly suffer two types of injury. Either they are bitten or clawed by the bird, or the bird pokes them in the eye with its beak.

**If poked in the eye by a bird's beak**:
Lie the casualty flat, with the head immobilised and slightly elevated. Instruct the casualty to close both eyes and to try not to blink. A clean dressing should be taped over both eyes. Medical aid should be sought.

**If bitten or clawed by the bird:**
Wash your hands, then clean the wound with antiseptic. If the wound is continuing to bleed, apply direct pressure with the fingers and elevate the wounded part. Apply a clean dressing or pad. A tetanus needle is advisable. If bleeding persists, medical help should be sought. If the wound becomes pussy, swollen, inflamed, or does not begin to heal after 48 hours, medical advice should also be sought.

## Lizards

**If bitten by a blue-tongued lizard or a goanna:**
Though the bites of these animals are not venomous, they may become infected, especially if the lizard has recently eaten carrion. The lizard may attempt to hold on to the victim with its jaws for some time before letting go. It is unwise to forcefully free the bitten area, as it may result in a more severe injury.

If the lizard is a blue-tongue, sit it out if possible, as it will eventually let go. Otherwise try holding a lit match or turned-on water hose to its mouth.

If it is a goanna, try the lit match method, or, if desperate, wedge a strong object such as a crowbar in its jaw, and force open. This may work.

**For superficial wounds**, clean the wound with antiseptic and apply a clean pad or dressing. A tetanus needle is advisable. If the wound becomes pussy, swollen or inflamed, or does not begin to heal after 48 hours, seek medical advice.

**If the wound is major**, it can become infected, particularly if the lizard has previously eaten carrion. Follow the procedure for superficial wounds, and seek medical advice.

It is a fallacy that wounds caused by these lizards recur or re-open every year or regularly. If the infection is severe, it may re-open because the skin has healed, but the infection underneath has not. If the infection is severe, it may be a long and troublesome recovery.

**If scratched or bitten by other lizards, drawing blood:**
Wash your hands and the injured area. Clean the wound with antiseptic. Apply a clean pad or dressing. A tetanus needle is advisable. If the wound becomes pussy, swollen, inflamed, or does not begin to heal after 48 hours, seek medical advice.

## Tortoises

**If scratched or bitten by a tortoise, drawing blood:**
Wash your hands and the injured area. Clean the wound with antiseptic. Apply a clean pad or dressing. A tetanus needle is advisable. If the wound becomes pussy, swollen, inflamed, or does not begin to heal after 48 hours, seek medical advice.

# Snake bites

If uncertain of the species of snake apply procedure as for a venomous snake bite. Snake bites occur mostly as a result of treading on the snake, aggravating it or handling it incorrectly. Wounds are usually sustained on the limbs. Sometimes very little venom is injected, even if fangs have penetrated the skin. However, first aid procedure should be carried out.

**Venomous Snakes**
The old method of cutting and sucking the bitten area is no longer considered effective first aid. Research has shown that firm pressure applied over the bitten area significantly delays the movement of venom. When pressure is combined with immobilising the limb, the chances of venom reaching the bloodstream is minimised.

**Symptoms**
Any of the following symptoms may appear within 15 minutes to two hours after the bite: drowsiness and fainting, headache, disturbed vision, nausea or vomiting, breathing difficulty or chest pain, collapse.

**What to do**
Do not attempt to capture or kill the snake as you risk being bitten. If possible, take note of its colour and size; otherwise the species can be determined by a Venom Detector Kit (used in most hospitals).

Keep calm, keep reassuring the victim and make the victim comfortable. Do not wash the venom from the bite. Remove any jewellery if bitten on the arm or hand.

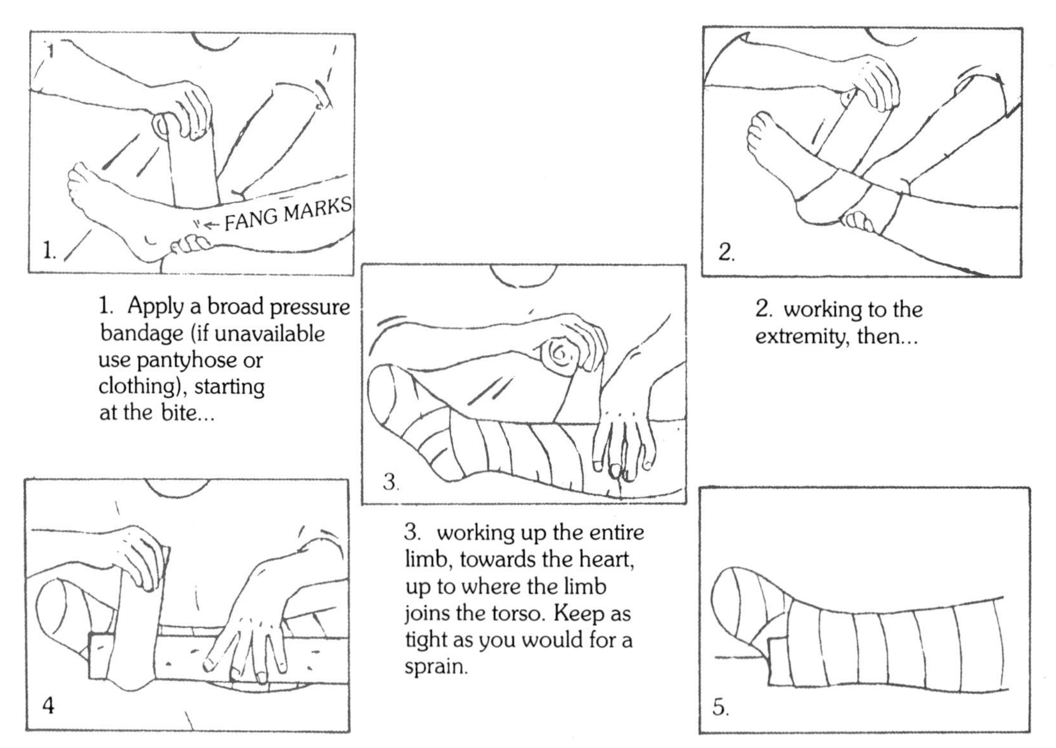


1. Apply a broad pressure bandage (if unavailable use pantyhose or clothing), starting at the bite...

2. working to the extremity, then...

3. working up the entire limb, towards the heart, up to where the limb joins the torso. Keep as tight as you would for a sprain.

4/5. Immobilise the limb. If it is a leg, splint to a stick. If it is an arm, put it in a sling. Do not attempt to remove clothing.

**If the bite is to the face**
It has been suggested that for an area where a pressure bandage cannot be applied, such as the face, pressure applied with fingers to the bite may help to slow the flow of venom. Do not remove fingers until help arrives. If there are three people present, leave one with the victim, send the other for help. If there are only two people present, the victim should be piggybacked to help, otherwise position the victim comfortably and seek help quickly. If the victim is alone in the bush he/she must apply the bandage, immobilise the limb and call out for help or walk slowly to the nearest point of help. If that is not possible, lighting a fire may help attract someone's attention.

**If bitten by a tree snake or python**, you will sustain puncture marks, bleeding from the punctured wound, and some people may suffer shock.

Clean the wound with antiseptic. If bleeding persists, apply direct pressure with your fingers. Apply a clean dressing or pad. A tetanus needle is advisable. If the wound becomes pussy, swollen, inflamed, or does not begin to heal after 48 hours, seek medical advice.

If the victim is suffering from shock: Lie the victim down, loosen clothing and make comfortable. Maintain body warmth. Keep reassuring. Do not administer any fluids or food. If concerned, or the victim shows no sign of improvement after 30 minutes, seek medical aid.

Note: If at any time after being bitten, the casualty experiences drowsiness or fainting, headache, disturbed vision, nausea or vomiting, breathing difficulty or chest pain, or collapses, apply procedures for venomous snakes and seek medical aid.

## Bee stings

Only female bees are capable of stinging, but only once. They inject a venomous barb into the skin and then die with their first sting. Most stings result from the bee being trapped, such as getting tangled in clothing or hair, or under foot.

**Normal reactions** to being stung result in local swelling and redness, and pain with possible itching to the bite area. The first thing to do is to remove the barb as soon as possible; scrape off by a scratching motion of the nails. Do not squeeze, as more toxin may inject into the wound. Wipe the area clean with water. Applying an ice pack to the bite may relieve the pain. Calamine lotion, steroid cream or other commercial insect-bite preparations can be applied to the bite area. If the victim's condition deteriorates or symptoms other than swelling and local pain appear, seek medical aid immediately.

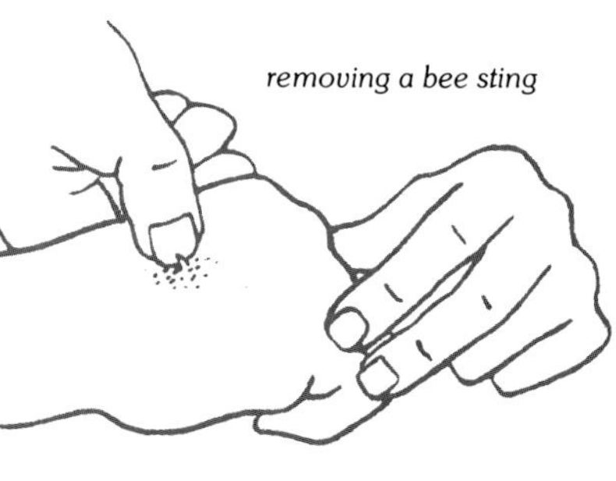

*removing a bee sting*

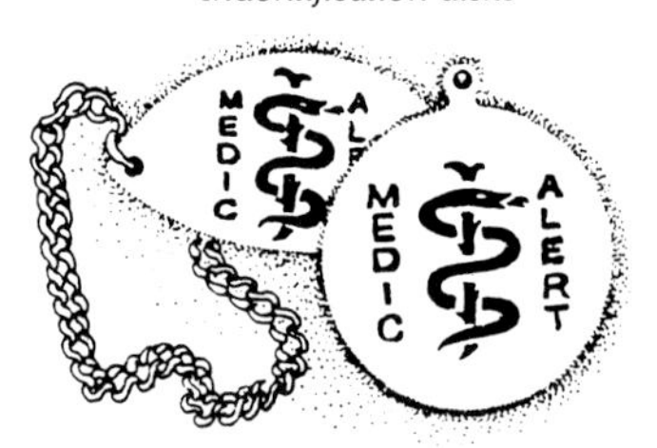


*Indentification disks*

**For people with allergies to bee stings**, a sting can be a life-threatening situation. Those people should identify themselves by wearing an indentification disk, advising of medication to be used if stung by a bee, and where it is kept.

After being stung, any of the following symptoms may appear in people with an allergy to bee stings: an irritating rash to neck and chest; swelling of the face and throat; breathing difficulty; wheezing; sneezing; nausea or vomiting; headache; disturbed vision; shock (feeling weak, faint or giddy, feeling cold and clammy, looking pale, shallow and rapid breathing). Unconsciousness may develop. If three people are present, one should get help, the other should apply the following first aid and stay with the victim. If two people are present, call for help immediately after applying first aid.

Keep the casualty at rest and keep reassuring. Remove the sting without squeezing it, as more toxin may enter the wound. Scrape off with a scratching motion of the nails. If the victim has medication for stings and is conscious to advise of the correct dose, help to administer it.

Apply a broad bandage to the bite area. Bandage from the sting, working up the entire limb towards the heart. Apply firmly but not so firmly as to restrict blood flow. Cover as much of the limb as possible. If bandage is not available, use panty hose or improvise with clothing. Immobilise the limb. If a leg, splint to a stick. If an arm, put in a sling. For bandaging technique, see SNAKE BITES page 12. If breathing stops, give mouth-to-mouth resuscitation. If the heart stops, start cardio-pulmonary resuscitation. Seek medical help immediately. It is preferable not to move the victim, but to get help to the victim.

**Note: Bee stings to the face or the neck** are potentially dangerous, as they can result in a swelling of the breathing passage and restrict breathing. Remove the sting as above. If the victim is conscious, give ice to suck and apply a cold compress to the bite area. Maintain an open airway. Position the victim in the most comfortable position (sitting position is the best), and loosen tight clothing. Transfer to a hospital as soon as possible, whether or not the victim has a known allergy to bee stings.

## Wasp stings

Only the female is capable of stinging. Unlike bees, wasps do not die after stinging, and are capable of stinging repeatedly without leaving a barb, though their toxin is not as toxic as that of bees.

**Normal reaction to wasp stings** is the same as that for bees, see BEE STINGS page 14. The treatment is identical, except that there will be no sting to remove in this case. For people with allergies to wasp stings, or for people who are stung about the face and neck, comments and treatment are the same as for bees, see BEE STINGS page 14.

European wasps are more aggressive than bees, and will attack if threatened or their nests are disturbed. They are capable of multiple stings, each one more painful than a bee sting, and without a barb.

**Normal reaction to being stung** results in local swelling and redness, which may be several centimetres in diameter, and severe pain with possible itching. Treatment of normal reactions is the same as that for bees, see FIRST AID, BEE STINGS page 14. There will be no sting to remove. If the victim's condition deteriorates, or symptoms other than swelling and local pain appear, the same procedures as for bee sting allergies should be adopted. See BEE STINGS page 14. Treatment for wasp stings to the face or neck is also the same as for bee stings.

# SPIDERS

If bitten by a spider take note of the spider's size, colour, and type of web (if any), as it is useful for its identification. If the offending spider can be caught safely it should be captured for identification.

## Funnelweb Spider bites

Bites from the male funnelweb are more toxic than those of the female.

Any of the following symptoms may appear anytime within ten minutes after being bitten: intense pain to the bite area; nausea and vomiting; abdominal pain; flow of perspiration; breathing difficulty; excessive salivation; weeping eyes; high pulse rate; collapsing.

1. **Keep calm and keep reassuring the victim**. Make the victim comfortable, loosen tight clothing. Do not give the victim fluids.
2. If the bite is on an upper or lower limb, **apply a broad pressure bandage, starting at the bite**. Bandage from the bite down to fingers or toes, then work up the entire limb towards the heart. Apply firmly enough to compress tissue, but not so firmly as to restrict flow of blood. Cover as much of the limb as possible. If bandage is not available, improvise with pantyhose or with strips of torn material.
   For bandaging technique see SNAKE BITES page 12.

3. **Immobilise the limb**, splint to a stick if a leg, put in a sling if an arm.
4. **If bitten on the face**, apply light pressure with your fingers to the bite area.
5. **If three people are present**, one should stay with the victim, while the other goes for help. If two people are present, position the victim comfortably and seek help quickly.
6. If you are bitten when alone in the bush, don't panic. If possible, apply the pressure bandage, and walk slowly to the nearest point of help. If that is not possible, lighting a fire may help to attract attention. If this can be done safely.

## Redback Spider bites

Only the female of the species is capable of inflicting a serious bite. Any of the following symptoms may appear after being bitten: pain at the bite site, becoming general; nausea; vomiting; flow of perspiration, sometimes profuse sweating; dizziness; faintness; chills. Take note of the spider's size, colour, and type of web (if any), as those points are useful for identification. Capture the spider if it can be done safely.

If someone is bitten: keep reassuring the victim. Loosen tight clothing. Apply a cold compress to the bitten area to relieve the pain. Do not freeze the tissue. Do not apply a constricting bandage. Seek medical aid.

## Brown Trapdoor Spider bites

Though this spider will strike if provoked, no deaths have been recorded as the result of bites. It is not regarded as dangerous and no control measures are necessary if bitten. The bite area can be quite painful, and applying a cold compress may relieve some of the discomfort. If in doubt about the spider's identity, it is wiser to treat as venomous, and apply procedures as for FUNNELWEB SPIDER page 15. Also seek medical help if the victim's condition deteriorates, or you are concerned.

## Black House Spider bites

Though bites are not fatal, the female of the species can inflict a painful bite. Shortly after the bite, any of the following symptoms may appear: severe local swelling; local pain; sweating; vomiting; shivering; weakness. Though no special first aid procedures are required, a cold compress to the bite area may relieve some of the pain. Should the victim be concerned, the victim's condition deteriorate, or the spider not be positively identified, seek medical aid.

## Huntsman Spider bites

The bites of some huntsman spiders are more toxic than others and some bites can cause considerable pain. The pain can be relieved by applying a cold compress to the bite area. People with various allergies to insect bites may show a greater reaction to the bite. If concerned, if the victim shows symptoms other than pain to the bite area, or the victim's condition deteriorates, seek medical help.

# MAMMALS

Most mammals covered in this book are marsupials. A marsupial is an animal with a pouch, whose young leave the birth canal in an incomplete, foetal state. Though tiny, blind and hairless, the young's well-developed sense of smell and strong limbs instinctively guide it to climb through the 'hairy forest' towards a nipple for food. Having reached the nipple, the young attaches itself firmly to a teat until its body develops sufficiently for life in the outside world. Placental mammals, such as bats, rodents, and of course human beings, nurture their foetus by means of the placenta. Monotremes, such as the echidna and the platypus, differ from other mammals because they lay eggs and lack nipples.

## Injured mammals

Mammals can sustain injuries from dog or cat attacks, from contact with moving vehicles, from man-made structures, such as fencing or electrical wires, or even from extreme weather conditions. Be cautious when dealing with an injured animal. Although some injuries may make it less mobile, it is likely to be more aggressive. Prior to treating the animal, aim to immobilise its defences i.e. cover the muzzles of animals that bite, and the claws of animals which claw. Some animals need special consideration: kangaroos' tails and feet, for example, must be given a wide berth.

If the animal is bleeding externally, control the bleeding by placing a rag over the wound, pressing lightly. Wrap the animal in a blanket, towel or cloth, and take it to a vet. If there is no external bleeding and it is possible to do so, examine the animal for signs of injuries, such as puncture marks or fractures. If there are no signs of injuries but the animal looks groggy, treat for shock, see TREATMENT FOR SHOCK page 17 and observe. In about an hour, assess the situation again to decide whether to release the animal or take it to a vet for examination. Handle the animal minimally.

You can offer the animal a solution of one teaspoon of glucose powder to a cup of water. If the condition deteriorates, or you are concerned, take the animal to a vet. For methods of transport, see individual entries. If you are satisfied with the recovery, release the animal in the area where it was found, see RELEASING MAMMALS page 20. **Note: With any injuries, maintaining body temperature is essential for the animal's recovery**. Wash your hands after handling animals.

## Treatment for shock

Treatment for shock is used to pacify the animal or settle it down. The procedure can be applied to almost all injuries, or if the animal is distressed, prior to examining or treating it. It is based on three objectives: noise reduction, maintaining warmth, and light reduction.

Find a container big enough to accommodate the animal comfortably. Line the bottom with cloth for added warmth, ensuring the animal does not become

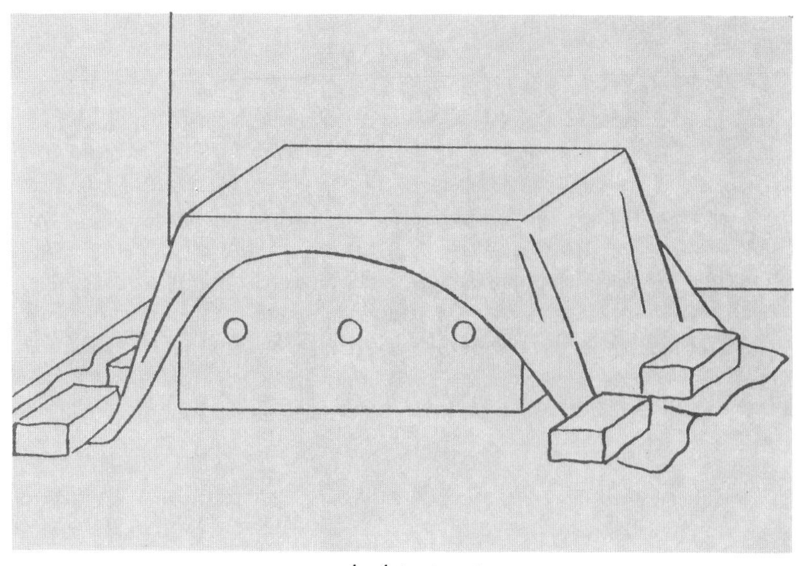

*shock treatment*

entangled. If airtight, punch holes in the sides or the lid for air-flow. Place the animal inside the box, and cover with the lid or cloth. (If using cloth, secure so it cannot be lifted.) That stops the animal from escaping and eliminates light. Place the box in a quiet spot away from excessive noise or vibration. On a cold day, place the box in a warm room if possible. Leave the animal alone for at least an hour. Examine or disturb as little as possible, and only when the animal appears to have settled.

Having seen some improvement in the animal, provide a source of energy. The most readily available one is a mixture of water and glucose powder. Mix at a ratio of a teaspoon of glucose to a cup of water. Offer the liquid in a container. Wet the animal's mouth and wait for it to drink. Do not force if the animal does not respond. When treating for shock, use your common-sense to create a suitable environment. A box may not be available, treatment may need to be carried out while travelling in a car, or the animal may be too big for any container. Always keep in mind the three necessities: minimal noise and light, and added warmth. Wash your hands after handling animals.

## Orphaned mammals

Upon finding an orphaned young, the first thing to do is to ensure that it is kept warm and placed in a quiet place. Depending on the

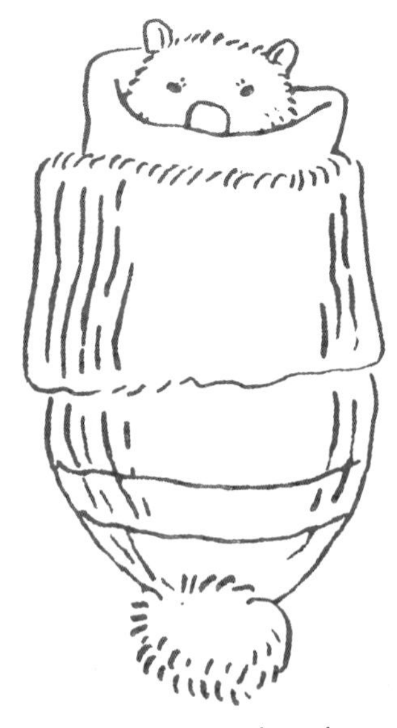

*improvised pouch*

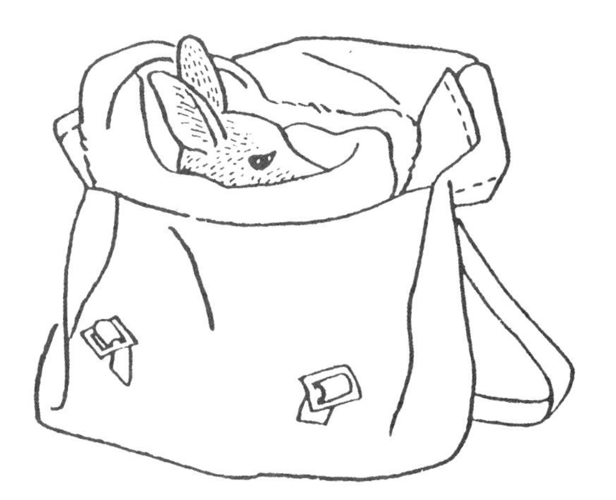

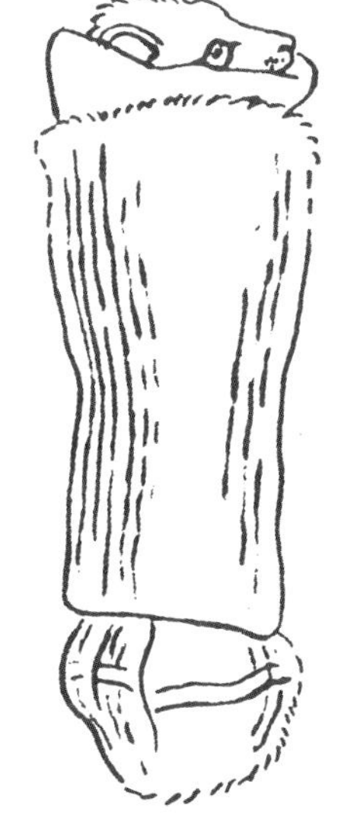

*improvised pouch*

species, that may involve improvising a pouch, for marsupials, or wrapping in a cloth and placing or hanging upside-down, for bats. In an emergency, the warmest place for a tiny unfurred animal is in contact with the warmth of the human body. As soon as possible, expert advice should be sought from someone who can rear the animal, or advise on animal rearing. See EXTRA HELP page 193. Also refer to REARING MAMMALS page 21. Depending on how long before expert help is obtainable, the animal may need to be fed. See page appropriate to the species for suggested food for the young, and FEEDING BABY MAMMALS page 20. If the baby has been found in bad weather conditions or as the result of an injury to the mother, take it to a vet for examination as it may have been injured or be in poor health. Wash your hands after handling animals.

## Mammals caught in electricity wires

Animals are safe on electricity wires, provided they can balance on one wire. If they come in contact with another wire, be it parallel to the one they are on, or one on the post, a current will pass through their body, causing instant death. However, if the animal is nursing a young one, the baby is not necessarily electrocuted with the parent. From a safe distance, assess if the animal is alive. If so, do not attempt a rescue, no matter how accessible it is. Call someone experienced in dealing with electricity, see EXTRA HELP page 193. Be prepared when help arrives. Have a container ready to accommodate the animal, and be prepared to take the animal to a vet if it appears injured or sick. If the electrocuted animal was nursing young, you may have an orphaned mammal to cope with. See ORPHANED MAMMALS page 18. Wash your hands after handling animals.

## Releasing mammals after captivity

Releasing an animal is not just letting it go, but releasing into a suitable environment. Prior to releasing, learn about the species, its natural habitat and its habits. The best place to release is in the area where the animal was found, exception being when the injuries it sustained there are likely to happen again. Consider the following when releasing: is the animal mature and well enough to fend for itself, or does it need further fostering? Does it help itself to food, or does it rely on humans to feed it? It may need rehabilitation prior to release. Is it territorial, or is it likely to risk attack by intruding on another animal's territory? Is it nocturnal or diurnal? Nocturnal animals should be released at dusk or night, diurnal in the morning. Are weather conditions suitable; should you hold onto the animal for another day? Is the area where you are considering releasing it suitable for its habitat? Does it have enough trees for shelter, or is the soil suitable for burrowing? Is the area too noisy; is there too much traffic; are there too many people around? Do not release in a National Park without prior consent from your State Fauna Authority. They can also help in choosing a suitable release spot. See EXTRA HELP page 193. Wash your hands after handling animals.

## Feeding baby mammals

Feeding young or unfurred mammals is a demanding job, and expert advice should be sought. See EXTRA HELP page 193. Feeding utensils can be obtained from chemists, pet shops, vets or animal-rearing organisations, depending on requirements. Most baby mammals should accept food from a bottle with a pet-teat, the size of teat depending on the size and age of the animal. Some very young or unfurred mammals may need to be fed through an eye dropper or a syringe with a teat. It is of utmost importance that the animal be made warm, prior to feeding. For added security, wrap the animal in a cloth and nurse while feeding. (Bats need to be held with the head lower than the feet.) An unfurred mammal will need feeding approximately every two hours, increasing time between feeds as it gets older.

Sterilise all feeding equipment and keep the animal clean, wiping the mouth area and any food spills off the animal. After feeding, rub the anal area with a tissue or soft cloth, to stimulate bowel movement. The animal should

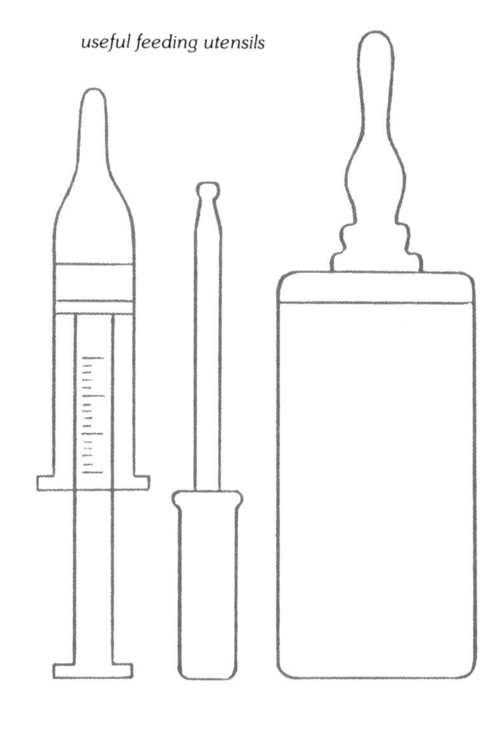
*useful feeding utensils*

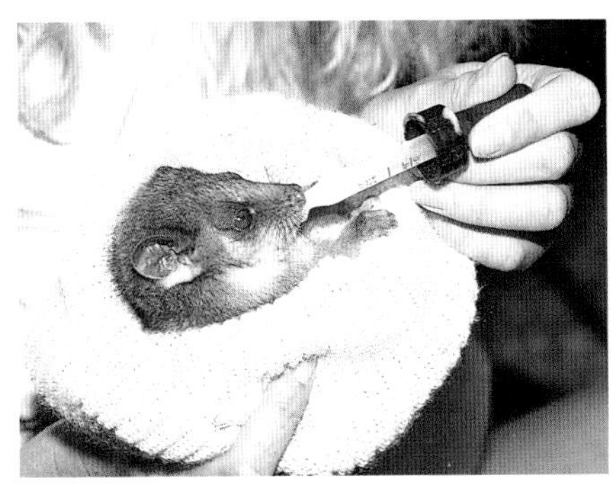

oblige instantly. Wash your hands after handling animals. Note: Feeding equipment suggested above does not apply to baby echidnas, which need to be tube-fed, see ECHIDNAS, Emergency food, page 47.

## Rearing mammals

The aim and information in this book is for emergency situations and short-term care of injured, orphaned and sick animals. However, when long-term care is necessary, keep the following in mind: a licence to hold a native animal should be obtained from your State Fauna Authority. See EXTRA HELP page 193. Some animals are extremely difficult to rear and their mortality rate in captivity is high. Koalas should be reared only by people experienced in the field of koalas. Unfurred and small mammals also need expertise in rearing, and echidnas require special feeding utensils due to the shape of their mouth. There is no such thing as a perfect formula for rearing mammals. This book suggests a few which have been recommended by various animal parks, vets and animal organisations. Undoubtedly, there are many more foods and formulas which can be used successfully. Some have been formulated especially for rearing animals (such as Wombaroo formula), and are available through some animal-oriented organisations.

When housing the animal in captivity, try to simulate its natural environment as closely as possible. Research the habitat of the species, natural food, and its habits. Some animals are social, and do better in captivity when with or near animals of the same species; others need to be alone. House away from excessive noise, and avoid changing the environment by carrying the animal around.

Apart from feeding and pottying the animal, there are other chores when rearing. Unfurred marsupials need to be rubbed with baby oil twice weekly, to stop their skin from drying out. Flying-foxes may need to be washed, as well as being rubbed with the oil. While caring for the animal, do not use it as a source of amusement or an exhibit to inquisitive people; nor should you make a pet of the animal. Keep in mind that one day it will need to survive in the wild. Should you be unable to cope with the rearing or need advice, see EXTRA HELP page 193, for a list of people who may be able to help. Wash your hands after handling animals.

## Poisoned mammals

It is not easy for an unexperienced person to diagnose that an animal is suffering from poisoning. If a group of victims in one area are suffering with the same symptoms, then poisoning may be considered as a possibility. Animals may be poisoned by consuming poison-sprayed plants, snail baits, rodent baits, industrial waste products, household chemicals, or through malicious deeds. Any of the following may be a sign of poisoning: nervous twitching or convulsions, fits, extreme salivation or frothing, symptoms of shock, difficulty in breathing, stumbling or loss of balance, or vomiting. Poison may result in internal haemorrhaging and an animal suspected of poisoning should be taken to a vet immediately. If malicious poisoning is suspected, report to your State Fauna Authority.

## Dead mammals on the road

A number of animals are casualties of vehicles on the road. Some of these are marsupials, pouched mammals. Examples of marsupials are: kangaroos, koalas, wombats, possums and bandicoots. Wombats, koalas and bandicoots have rear-opening pouches, which are set lower on the body than those of kangaroos.

Death to the parent does not always mean death to the baby. When you see a dead marsupial on or by the road, the following procedure may save a young one. If the animal is still on the road, remove it to the side. Push with a stick if you do not wish to handle it, keeping your safety in mind and avoiding the traffic. Turn the animal to the side to examine if it has a pouch. Wearing gloves or having wrapped a cloth around your hand, open the pouch and examine for a young one. If the animal is stiff, it may be necessary to force or tear the pouch open, as a stiff parent can still have a live young inside. If there is a baby inside, gently remove it and wrap in a cloth to maintain body temperature. See ORPHANED MAMMALS page 18.

Unfurred joeys at early stages of life are firmly attached to the teat. When taken off, they may tear at the mouth. At this early stage of life, it is difficult to save them. Should the pouch be empty but have an evidently suckled teat (reddish or swollen), examine the area for a wandering baby. The distressed young may have wandered off. Should you find the young one, see ORPHANED MAMMALS page 18. Wash your hands after handling animals.

## Euthanasia

Euthansia is best and most humanely performed at the hands of a veterinary surgeon. Authorities from National Parks and Wildlife NSW advise that it is illegal for the public to put down native animals. However, there may be extreme circumstances which would create an exception. If the animal is suffering, the condition is irreparable and help is not available, it may be more humane to kill the animal on the spot. People found abusing this exception will be penalised. Do not put down adult female marsupials, without first checking the pouch for a baby. Depending on the size of the animal and the tools available, any of the following are acceptable methods.

Place on soft dirt and put a sharp object, such as a spade, through the neck. Grab the animal by the hind limbs, and strike the head on a hard surface, such as a rock. Strike a heavy blow to the back of the head with a hammer or a heavy object. Place the animal in a bag, and drive over it with the car. Place the animal into a bag, into which you have punched several holes.

# POSSUMS

Tie securely to the outlet of the car exhaust pipe and start the engine. Let the engine run for a few minutes to allow the fumes to kill the animal. Shoot at the back of the neck with a firearm. Slit the throat, ensuring that the spinal cord is severed.

Before proceeding to put down the animal, give yourself one last chance to assess the situation. Is euthanasia the only solution? Can this animal still be saved at the hands of an expert? Is there no vet anywhere within reach? Could someone else deliver the animal to a vet? Having decided to put down the animal it is your moral obligation to ensure the job is done with minimal pain to the animal and to the point where the animal is dead. Wash your hands after handling animals.

Important note: These procedures have been written in the humane interest of minimising the suffering of mammals. Remember, there are heavy fines for any abuse.

## How to deter

Possums which insist on nesting in roof cavities can be a problem. Noise made by the possums can disrupt sleep, and the excreta can eventually stain ceilings. Possums can be deterred from entering roofs by various methods. Trees, particularly ones overhanging the roof, are a means of climbing onto the roof and into the cavity. Cut off any overhanging branches or fit a cylinder over the base of the tree. The cylinder must be made of non-climbable material, such as metal.

Carefully examine the roof to determine where the possum may be entering. Seal off any possible entry points, no matter how small. If it is not possible to stop possums from entering the roof, try making their stay an unpleasant one. The smell of camphor or naphthalene sprinkled in the area, or burning mosquito coils, makes the place uninviting to a possum. Too much light in the area where they wish to nest is another deterrent. Leave a light on in the roof area for three consecutive nights. If using an exposed bulb, ensure it does not touch any surface, as it may start a

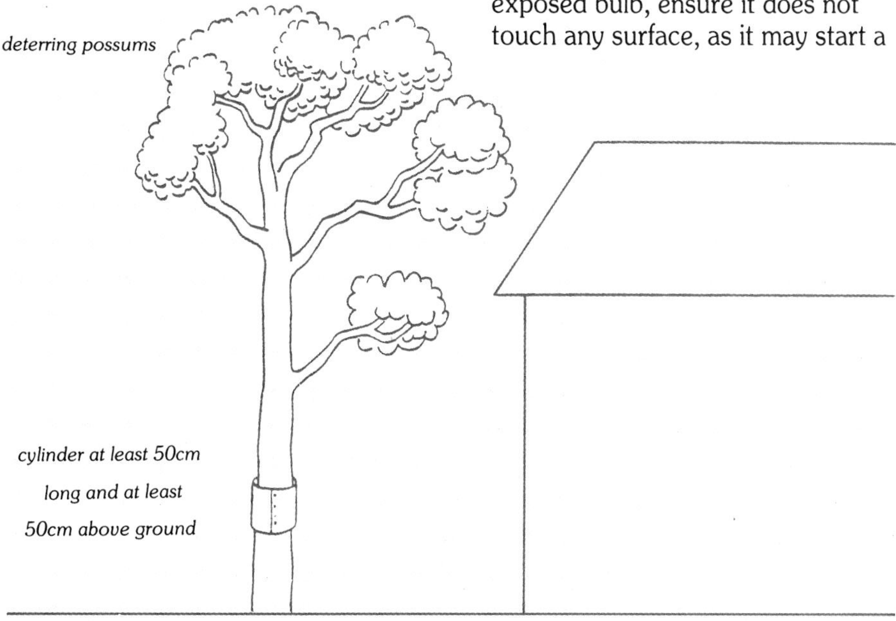


*deterring possums*

fire. **Note: When sealing the roof from possums, ensure that there are no possums in the roof at the time.**

To deter possums from nibbling plants, a spray can be used. Boil a packet of Quassia chips (available from a chemist) in ten litres of water for two hours. Strain, discarding the chips and reserving the liquid. Add a further 20 litres of water to the liquid. Spray the affected area for five consecutive days, and again after rain, until the problem is eliminated. Another product for this purpose is D-TER. It is a powder and is available from some hardware stores. Use as per instructions on the pack.

*deterring possums*

*chop off any connecting branches*

## Trapping possums

There are a few options in dealing with a possum problem. You can call out a reliable pest control company to deal with the problem. For a fee, major ones will trap the possum, release it elsewhere, and seal off the suspected entry point/s. You may wish to try to deter the possum from nesting on your property, and encourage it to find alternative accomodation. See POSSUMS HOW TO DETER page 23. Or you may wish to trap the possum yourself. Before trapping, obtain a licence from your State Fauna Authority, to handle a protected native animal. You can hire the trap, see EXTRA HELP page 193, or you can make the trap according to the diagram shown.

Before evening, place the trap in the roof area, or near the point where the possum may be entering. Place a bait in the trap. Use bread

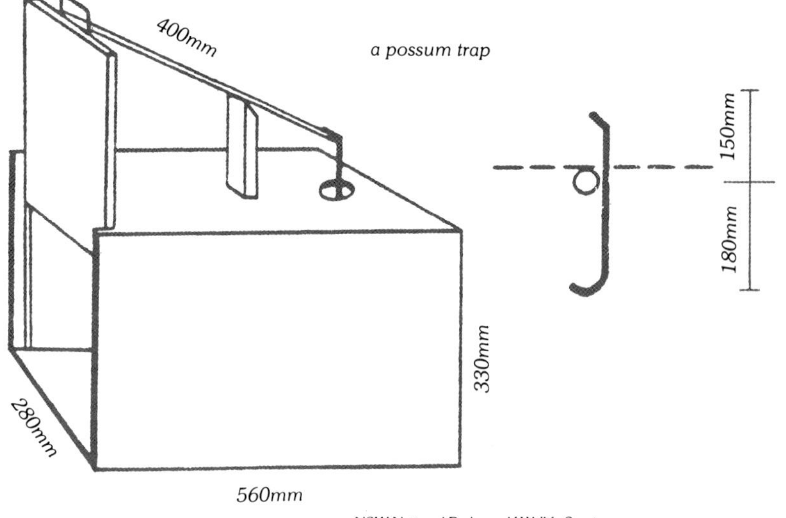


*a possum trap*

NSW National Parks and Wildlife Service

with honey, chopped fruit, or a rag sprinkled with aniseed liquid (available from a chemist) or aniseed base liqueur. Set the trap. Listen for noises around the trap, or check the trap every evening.

If the possum has been trapped, it is your responsibility to release it. Once the possum has been trapped, check the roof area for any nesting material. Destroy the nest and spray the area with a domestic insect surface spray. It is most important to block off any points of entry used by the possum so that the problem is not repeated. For the next few evenings, listen for noises in the roof to ensure that possums, possibly babies, have not been left behind. It would be inhumane to leave them in the sealed roof. Wash your hands after handling the possum or the nest.

**Releasing trapped possums**
To release the possum in a National Park, prior consent from your State Fauna Authority will be needed. They can also advise you of possible release spots. See EXTRA HELP page 193. Release the possum after dark or in the evening, as it is a nocturnal animal. If the weather conditions are extremely bad, consider holding on to the animal till the next day. Choose a natural bushland setting, away from traffic, and at least eight kilometres from the site where it was trapped. Anything closer risks the possum returning to the area where it was trapped. If possible, provide the possum with a new home in the new area. Make a box (see HOW TO ATTRACT page 27) — though it need not be so elaborate — and attach to a tree at the release site. Place the bait used for trapping in the new box, and transfer the possum from the trap into the box. For handling the possum, see COMMON BRUSHTAIL POSSUM Handling page 28.

Consider releasing the possum in your yard, possum-proofing the house first. Remember to provide shelter for the possum, see HOW TO ATTRACT page 27.

Being territorial, the possum in your yard will deter other possums from entering the yard. By supplementing its diet, you minimise chances of the possum's desire to nibble and destroy your garden plants. Regularly put out chopped fruit on a feeding tray or at the base of a tree. Lock up your pets for the night, as they are potentially dangerous to nocturnal, native animals.

## Common Brushtail Possum *Trichosurus vulpecula*

**Identification**
Adult head and body length 35 to 55 cm. Tail length 25 to 40 cm. About the size of a domestic cat. Muzzle is pointed, ears long and conspicuous. Tail is covered in brushy fur except for a small area underneath. Fur colour varies regionally, in shades of grey, ginger and black. Albinos are common in Tasmania.

**Habitat**
Leads a solitary life, occupying a marked territory. Found in most areas with trees, such as forests and woodland. Spends the day in hollow limbs of trees, though can

Other names: Brushtail Possum, Silver-grey Possum, Long-eared Possum

Native. Marsupial. Territorial. Nocturnal.

shelter in fallen logs, on tree branches, on the ground or in rabbit burrows. Has adapted well to the urban environment, frequently taking up residence between ceilings and roofs of houses and in cavities within the home.

**Breeding**
Breeds at any time of the year, though mostly during colder months. Usually gives birth to a single, unfurred young. First four to five months are spent in the mother's pouch. The next one to two months are spent riding on the mother's back as the weaning process begins. By seven to nine months the young is fully independent, though still follows the parent for some time.

**Food**
A vegetarian. Eats a variety of leaves, eucalypt trees being the favourite for both the leaves and the flowers. Also feeds on native fruits, buds, bark, grass and an odd small insect. When frequenting

yards, has been known to scavenge on a variety of leftovers, including meat, with a particular liking for rosebuds and shoots of vines.

**Emergency food**
Adult: eucalypt leaves, especially young leaf tips; fruit such as apples, bananas, oranges, mandarines and grapes; vegetables such as capsicum and tomatoes; native flowers such as those of eucalypts, melaleucas and grevilleas; bread with honey. Short-term feeding: one teaspoon of glucose powder to a cup of water is the best emergency food for a sick possum. Baby: either evaporated milk diluted with warm water (ratio 1:1); or, one scoop Digestelact (milk powder from a chemist) to 100 ml warm water. Strengthen the formula as the animal gets older. For long-term feeding, add two drops of baby vitamin drops and a pinch of glucose powder to the formula. See FEEDING BABY MAMMALS page 20.

**How to attract**
It is wise to possum-proof your house (see POSSUMS HOW TO DETER page 23, before encouraging them to enter your yard. If possible, do not 'tidy' trees by chopping down dry branches, as they may contain hollows, which are potential nesting places for possums. If there are no trees or hollows nearby, construct a wooden box as shown in the picture, and mount on an upper branch of a tree. Do not place on a forked branch as that provides access for cats. Flowering natives such as eucalypts, grevilleas, native figs, bottlebrushes, melaleucas and flowers of umbrella trees are a

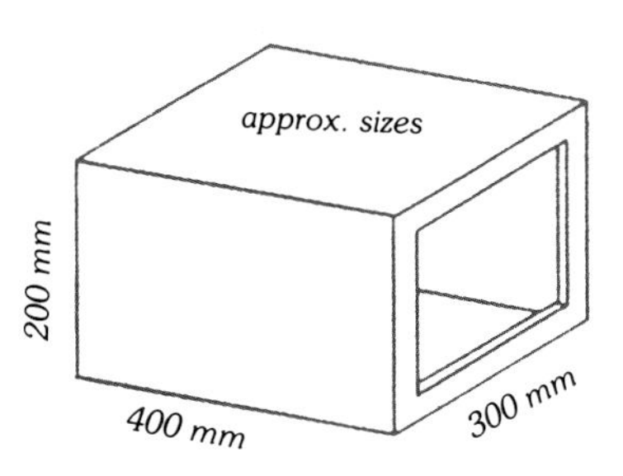


source of food for possums as well as other wildlife. Various cultivated fruits such as apples, bananas or oranges can be chopped and placed at base of trees or on a feeding tray, in the evening. Ensure all pets are locked up overnight, as they are potentially dangerous to nocturnal native animals. Be considerate to your neighbours. Keep in mind that the possums you set out to attract may end by taking up residence in your neighbour's yard.

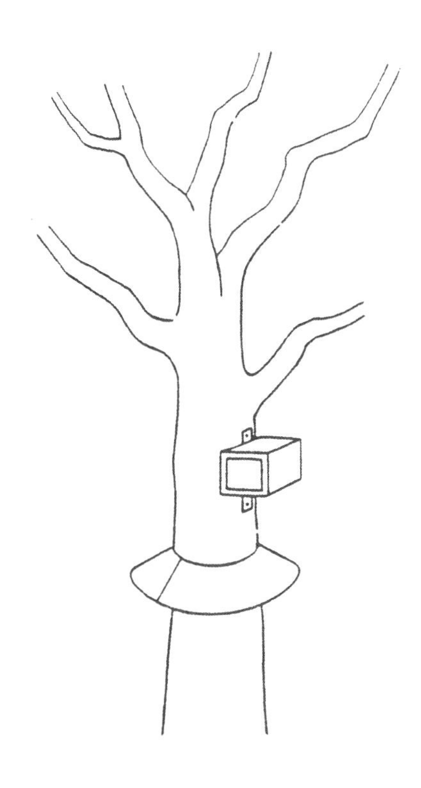

## Handling

Possums are capable of scratching and biting when handled, particularly if they are injured or have been trapped. Always have a transport/holding container ready prior to capture. Unless you're experienced, the easiest and safest way to handle a possum is through a towel or cloth. Place over the possum, covering the head. This disorientates and calms it. Scoop up with the cloth and lower into the waiting container, lifting the cloth off. Possums can also be handled by the tail. Hold away from your body, particularly your legs, and swing gently to stop it from climbing up its tail or your leg, and biting you. Place in the container head first. Another method of handling is to pick up behind the neck with one hand and at the base of the tail with the other. Heavy duty gloves are recommended for this method. Wash your hands after handling animals.

## Transport

Baby marsupials must be kept in improvised pouches for warmth and security. A sewn-up sleeve of a woollen jumper, a small container such as a basket lined with soft cloth, a sheepskin slipper, woollen hat, or if the animal is tiny, a woollen sock, are all suitable. Put a lining of soft material between the wool and the animal. Adults can be transported in bags made of strong but soft material such as pillowcases, or cardboard boxes. The box must have air-holes and a tight-fitting lid to stop the animal from escaping. It may be lined with straw but not material, as the animal may get entangled. If the animal is sick or injured, however, it may need immobilising and warmth, and may need to be wrapped in a blanket or towel and nursed while travelling. If transporting in a cage, cover with a cloth to create darkness. That calms the animal and makes it feel more secure.

### Did you know?

*Man's progress has not been kind to native animals. When man destroys the habitat of native animals, not only is their shelter eliminated, but their breeding is hindered and their food supply limited. Brushtail Possums are among the great survivors of man's progress. They have not only assimilated into the urban lifestyle, but utilised it to their advantage. Roofs of houses have made excellent dwelling places and foodscraps a quite acceptable diet.*

## **Common Ringtail Possum** *Pseudocheirus peregrinus*

Other names: Ringtail Possum

Native. Marsupial. Territorial. Nocturnal.

### Identification
Adult head and body length 30 to 35 cm. Tail length 30 to 35 cm. Smaller than brushtail possums. Muzzle is pointed, ears small and low set. Tail thin, slightly furred, with a white tip; the tail is often carried in a coil. Fur colour varies regionally, in shades of grey, ginger and black.

### Habitat
Leads a solitary life, though tends to be more sociable than the brushtail possum. Found in most areas with trees, favouring dense vegetation. In southern parts of Australia, builds a nest lined with grass or bark, in hollow limbs of trees or in dense undergrowth. Further north, spends the day in hollow limbs of trees, sometimes lining them with dry leaves. Rarely takes up residence in houses.

### Breeding
Breeds at any time of the year, though mostly during colder months. Usually gives birth to twins. Unfurred at birth, first four months are spent in the mother's pouch. By then fully furred, the weaning process begins. The next two months are spent riding on the mother's back or staying at the nest as she forages for food. By six months of age the young are fully weaned.

### Food
A vegetarian. Eats a variety of leaves, fruit and blossoms, eucalypt trees being favourite. Also feeds on bark, nectar and some seeds. Not partial to household scraps, unlike the brushtail possum.

### Emergency food
Adult: eucalypt leaves especially young leaf-tips; fruit such as apples, bananas, sultanas and grapes; native flowers such as those of eucalypts, grevilleas; bread with honey. Short-term feeding: one teaspoon of glucose powder to a cup of water is the best emergency food for a sick possum.
Baby: either evaporated milk diluted with warm water (ratio 1:1); or one scoop Digestelact (milk powder from a chemist) to 100 ml warm water. Strengthen the formula as the animal gets older. For long-term feeding, add two drops of baby vitamin drops and a pinch of glucose powder to the formula. See FEEDING BABY MAMMALS page 20.

### How to attract
As Ringtail Possums are not likely to take up residence in your roof, possum-proofing your house is not necessary before encouraging them to enter your yard. They are more likely to enter yards which are rich in food and shelter. Shelter can be provided by tree foliage, and dense trees such as melaleucas are useful. Natives such as eucalypts, grevilleas, native figs, bottlebrushes and flowers of umbrella trees are good sources of food. Ensure that pets are locked up for the night, as they are potentially dangerous to nocturnal native animals.

### Handling and transport
See COMMON BRUSHTAIL POSSUM page 28.

# GLIDERS

## Sugar Glider *Petaurus breviceps*

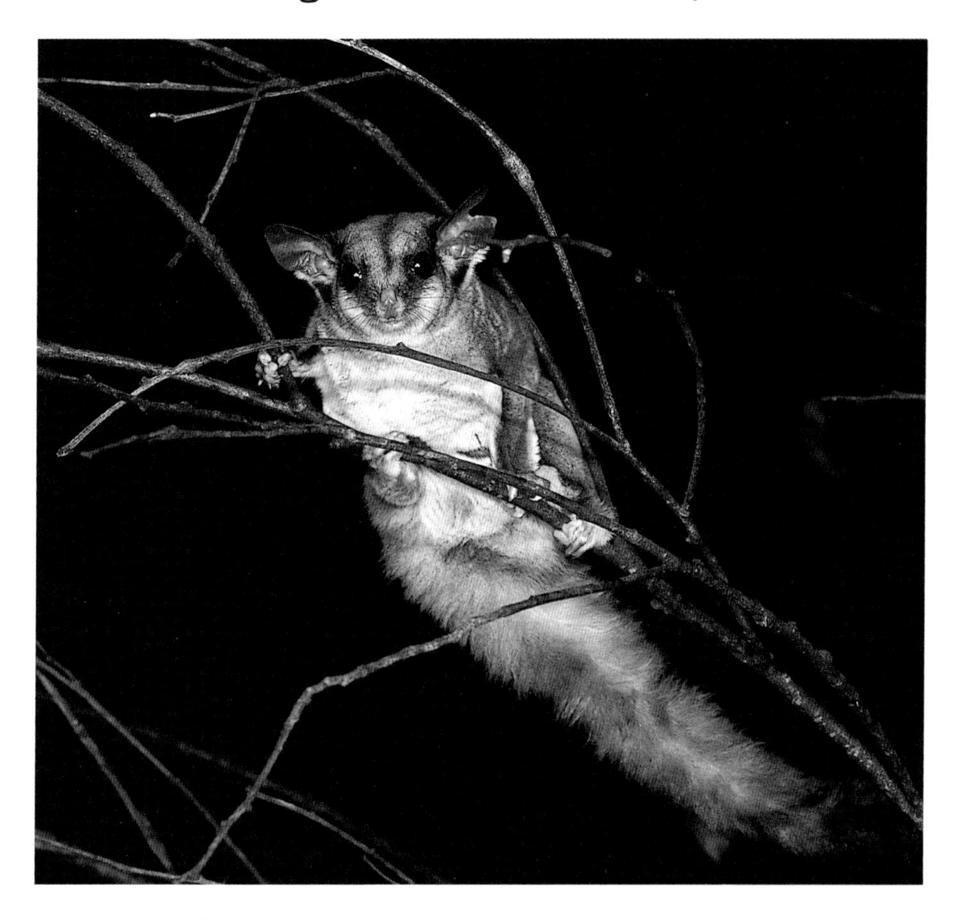

Other names: Squirrel Glider, Lesser Glider, Lesser Flying Squirrel

Native. Marsupial. Territorial. Nocturnal.

### Identification

Adult head and body length 16 to 21 cm. Tail length similar to that of body. When sitting, looks like a possum. Wrist and ankle are joined by a membrane, which is more evident when the animal is gliding. Overall body colour is greyish, with a dark stripe running along the back, from the nose to the tail. The tail is brushy and frequently tipped white or black. Belly and membrane are generally lighter in colour than the body.

## Habitat
Lives in social groups of up to seven adults and their young. Found in forests and woodland, where families huddle in leaf-lined tree hollows. Frequents backyards where it is most commonly seen on eucalypt and acacia trees.

## Breeding
Breeds at various times of the year, though mainly in the colder months. Usually gives birth to twins. Unfurred at birth, the first ten weeks are spent in the mother's pouch, followed by a further month or so in the communal nest. Though weaned by 14 to 15 weeks, the young remain with the mother until leaving the nest and the group, at approximately seven to ten months of age.

## Food
A vegetarian and insect eater. Foods include gum produced by acacia trees, sap obtained from eucalypts by biting into the trunk, nectar, blossoms and a variety of insects such as beetles and moths.

## Emergency food
Adults: insects such as moths and beetles; mealworms (available from some bird breeders); honey and water (ratio 1:1); soft ripe fruit such as avocado, paw paw, bananas; blossoms from eucalypts, grevilleas, hakeas; and pieces of bread spread with jam or honey.
Short-term feeding: honey mixed with water (ratio 1:1) is the best emergency food for a sick Sugar Glider.
Baby: either evaporated milk diluted with warm water (ratio 1:1); or one scoop Digestelact (milk powder from a chemist) to 100 ml warm water. Strengthen the formula as the animal gets older. For long-term feeding, add two drops of baby vitamin drops and a pinch of glucose powder to the formula. See FEEDING BABY MAMMALS page 20.

## Handling
Gliders are capable of scratching and biting when handled. They may be more aggravated when injured, though some injuries may make them less mobile. Always have a transport/holding container ready prior to capture. When confronted, gliders can curl up in a ball, concealing their head within their body for security. Therefore it is best to create a dark environment as soon as possible to minimise stress. Unless you're experienced, the easiest and safest way to handle animals is through a towel or a cloth. Place over the animal, covering the head. Scoop up with the cloth and lower into the waiting container, lifting the cloth off. Gliders can also be handled by the tail. Hold away from your body, and swing gently to stop it from climbing up its tail or your leg and biting you. Place in the container head first. Another method of handling is to pick up behind the neck with one hand and at the base of the tail with the other. Gloves are recommended for this method. Wash your hands after handling animals.

## Transport
As for COMMON BRUSHTAIL POSSUM, TRANSPORT, see page 28

## How to attract
Tree hollows are potential nesting places for Sugar Gliders as well as other animals. Where possible, do not 'tidy' the trees by removing any

dead branches. The following natives are a source of food for gliders as well as other native animals. Choose ones best suited to your area. See EXTRA HELP page 193 for help in choosing. Acacias are a favourite, as well as bottlebrushes, flowering eucalypts, banksias, grevilleas, hakeas and xanthorrhea (grass trees). Ensure that all pets are locked up for the night, as they are potentially dangerous to nocturnal native animals.

**Did you know?**

*A Sugar Glider is capable of gliding over an area of up to 50 metres. The membrane tension is adjusted on either side for steering, and a tree on the 'flight' path acts as a brake.*

# KOALAS

## Koala *Phascolarctos cinereus*

Other names: Koala Bear, Native Bear

Native. Marsupial. Territorial. Nocturnal.

## Identification
Adult head and body length 70 to 82 cm. Generally, specimens found in the north of Australia are smaller and have thicker fur than ones further south. In Victoria, a Koala can weigh over ten kilograms, whereas in Queensland, Koalas average five to six kilos in weight. Fur is woolly and a light to dark shade of grey, mottled with white. Eyes are small, ears large (fringed in the north of Australia), nose large and naked.

## Habitat
Leads a solitary life. It is limited in area due to its food requirements, and found mainly in forests and woodland containing eucalypt trees. During most of the day, sleeps in a fork of a tree, descending only to move to another tree.

## Breeding
Breeds mainly in spring and summer. Usually gives birth to a single young. Unfurred, it climbs out of the birth canal and into the pouch, where it remains for up to seven months. The next four to six months are spent in and out of the pouch, and on the mother's back, as the weaning process begins. At 12 months, it is fully weaned and by 18 months, leaves the mother.

## Food
A vegetarian. Consumes one to two kilograms of leaves daily, depending on size of the animal. Majority of leaves are those from eucalypt trees, of which there are over 500 species. Koalas favour approximately 40 varieties of eucalypts, each animal showing a preference for a few of the 40 varieties. The preference also changes seasonally, and with age. Moisture is largely derived from the leaves, though Koalas have been known to drink when sick or during extreme drought. The Koala's digestive system is adapted to cope with gum leaves, which are hard for other animals to digest; some are toxic to other animals.

## Emergency food
Adult: all injured or sick adult Koalas, should be handed over to expert hands, as soon as possible. See EXTRA HELP page 193. Should it be necessary to feed a Koala in an emergency, keep in mind that koalas prefer certain species of eucalypt trees. Offer gum leaves, preferably from trees where the animal was found. Otherwise, yellow gum, manna gum, river red gum, messmate, candlebark, Gippsland grey box, sallow box and grey box are some eucalypt varieties consumed by Koalas. Expert advice should be sought to identify the tree and determine whether it is consumed by Koalas in that area. Leaves with a yellow tinge should be avoided, as they may be toxic. Make water available for sick Koalas. Short-term feeding, one teaspoon of glucose powder to a cup of water, is the best

emergency food for a sick Koala.
Baby: All injured, sick or orphaned Koalas, should be handed over to expert hands, as soon as possible. See EXTRA HELP page 193. In an emergency, feed either: evaporated milk diluted with warm water (ratio 1:1), or one scoop Digestelact (milk powder from a chemist) to 100 ml warm water. The formula can be strengthened for older animals. For prolonged feeding, add two drops of baby vitamin drops and a pinch of glucose powder to the formula. See FEEDING BABY MAMMALS page 20.

### Handling

Though cute and cuddly looking, Koalas can bite and scratch when being handled. Always have a transport/holding container ready prior to capture. Unless you're experienced, the best and safest way to handle animals is through a towel or a cloth. Place over the animal, covering the head. This disorientates and calms the animal. Scoop up with the cloth into the waiting container. Koalas can also be picked up by placing one hand behind the scruff of the neck, and the other at the rump. It is advisable to wear gloves or cover your hands with a cloth, when handling with this method. Wash your hands after handling the animal, and remember: **all injured, sick or orphaned Koalas should be handed over to expert hands, as soon as possible**. See EXTRA HELP page 193.

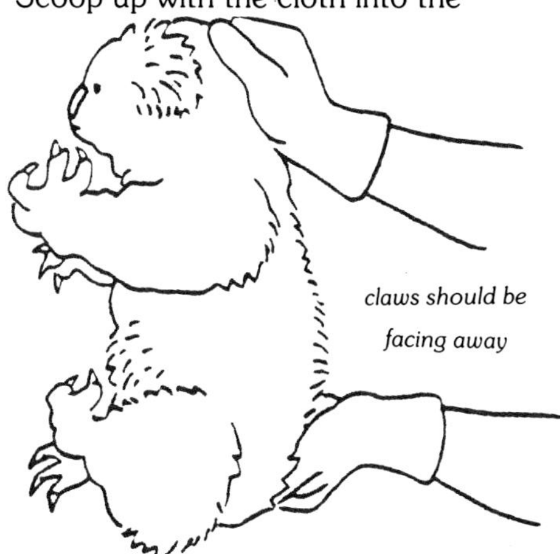


### Transport

For transporting Koalas, see COMMON BRUSHTAIL POSSUM, TRANSPORT page 28. You can offer the Koala a rolled towel to hug.

### How to attract

There are very few urban areas that can boast the presence of Koalas. However, we can all help in the preservation of this beautiful animal. In areas where Koalas occur naturally, particularly areas which are being cleared, new trees can be planted. Trees can also be planted to form corridors, joining various Koala feeding areas. Eucalypts are the predominant diet, however further research will be needed to determine which ones would suit the Koalas in a particular area. Should you be lucky enough to live in an area with a Koala population, lock your pets up for the night as they are potentially dangerous to native, nocturnal animals. (Your State Fauna Authority may be of more assistance on the subject.)

### Did you know?

*Koalas are not bears. Being pouched, they are marsupials. Bears do not have pouches. The Koala's closest relatives are the wombats.*

# KANGAROOS

## Western Grey Kangaroo *Macropus fuliginosus*

Other names: Black-faced Kangaroo, Stinker, Sooty Kangaroo, Mallee Kangaroo

### Identification
Adult stands to 1.5 metres, extended length head to tail being to 230 cm. Body colour varies in shades of grey, from silver through brown, to reddish. The fur is notably woolly.

### Habitat
Lives in groups. Has a wide distribution, ranging from grassland to forest areas. During midday, sleeps in shade of trees and shrubs. Grazes from late afternoon till morning.

### Breeding
Breeds throughout the year, though more in summer months. Usually gives birth to a single young. Unfurred at birth, the young climbs out of the birth canal, and journeys into the pouch where it attaches itself onto one of the four teats. By five to six months, it is fully furred and remains in the pouch till about 11 months of age, though in the last two months some time is spent outside the pouch as well. Though out of the pouch, it suckles till about 18 months old, by which time another newborn may have begun its life cycle.

### Food
A vegetarian. Grazes on grasses, from late afternoon till early morning. Also feeds on some shrubs.

### Emergency food
Adult: fresh grass; hay; small amounts of stock feed such as dairy cubes; bark for chewing; wholegrain bread. For older joeys, dry dog kibble; horse pellets; chopped fruit and vegetables. Short-term, one teaspoon of glucose powder to a cup of water is the best emergency food for a sick Kangaroo, Baby: either evaporated milk diluted with warm water (ratio 1:1); or one scoop Digestelact (milk powder from a chemist) to 100 ml warm water. Strengthen the formula as the animal gets older. For prolonged feeding, add two drops of baby vitamin drops and a pinch of glucose powder to the formula. See FEEDING BABY MAMMALS page 20. If the joey develops diarrhoea, stop the milk formula immediately. Offer glucose powder in warm water (ratio one teaspoon to a cup). Resume formula feeding when the diarrhoea has stopped. Take the animal to a vet if the diarrhoea does not stop within 24 hours of feeding glucose and withholding milk.

## **Eastern Grey Kangaroo** *Macropus giganteus*

Other names: Great Grey Kangaroo, Scrub Kangaroo

Native. Marsupial. Active by day and night (anytime but midday).

## Handling

There is no safe way to handle a large kangaroo, unless you're experienced. They are capable of biting, scratching, and injuring with their feet and tails. Smaller ones can be handled by the tail. Always have a transport/holding container ready, prior to capture. Without looking too obvious, grab the animal firmly by the tail, keeping the feet away from you. If placing into a sack, hold it above ground level to stop the animal from knocking its head on the hard surface. In a swinging motion, place the kangaroo into the waiting sack, head first. Once in a dark surrounding, the animal should settle.

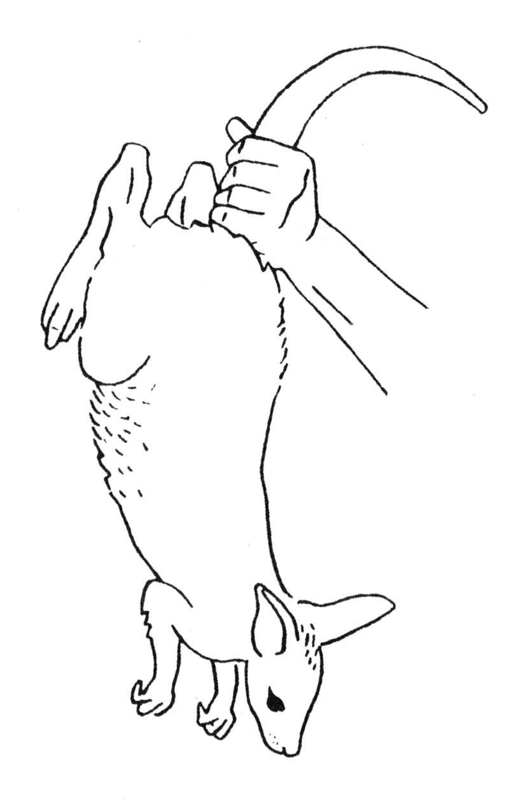

A small kangaroo can be handled by placing one hand under the chest close to the hind feet, and holding on to the base of the tail with the other. Keep the feet away from you. That method may also be applied to small kangaroos with injuries to their tails.

A smaller kangaroo can also be handled through a towel or cloth. Place over the animal, covering the head. That calms it and disorientates it. Scoop up with the cloth, keeping the legs free of pressure in case the animal kicks. An orphaned or injured kangaroo may need to be nursed in the cloth, while transporting. Wash your hands after handling animals. For injuries caused by kangaroos, see FIRST AID, MAMMALS, page 10.

### Transport

Baby and young marsupials must be kept in improvised pouches for warmth and security. A sewn-up woollen jumper, or a bag such as a field-bag from a disposal store and lined for warmth, are suitable. If using wool, put a lining of soft material between the wool and the animal. Older kangaroos can be transported in a sturdy, deep box lined with straw or hay, or a strong sack. While transporting in a car, the bag can be suspended. The animal can also hang suspended in a blanket or strong material. A sick, orphaned or injured animal may need to be wrapped in a towel or cloth for warmth and security, and nursed while transporting.

### How to attract

If living in an urban area, your chances of attracting a kangaroo into the yard are minimal, to say the least. However, in a rural environment, a paddock with shrubs and some native grasses may entice kangaroos to graze in the area. If you have a dog, keep it away from the area, both day and night.

### How to deter

Though kangaroos are responsible for damage to crops and pastures, they are not encountered in urban areas. Major problems should be referred to your State Fauna Authority, which runs a supervised culling program to combat the problem. They can also advise on special and electrical fencing which can be erected to help deter kangaroos. For problems on a smaller scale, a good strong, high fence around the area that needs protecting should keep kangaroos away.

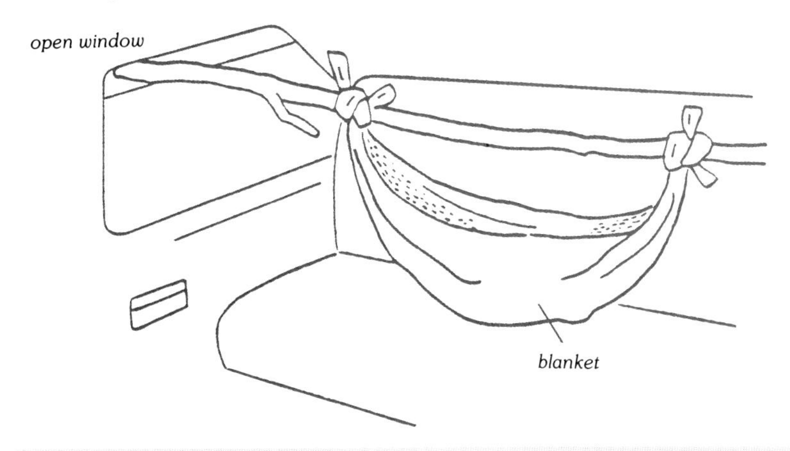


### Did you know?

*Kangaroos are capable of being 'permanently pregnant'. With one baby suckling on the teat, a youngster nearby that's not as yet weaned, chances are the mother is already pregnant again. As the pouch vacates, there usually is another infant ready to occupy it. Nature has truly equipped this animal for the task of motherhood. The two teats, feeding youngsters of different ages, produce two different strengths of milk.*

# WOMBATS

## Common Wombat *Vombatus ursinus*

Other names: Wombat, Naked-nosed Wombat

Native. Marsupial. Territorial. Nocturnal.

### Identification

Adult head and body length 85 to 115 cm, with a very short tail. Stout, powerfully built body; can weigh up to 30 kilograms. Head is broad, with small eyes. Ears are small and pointed and the nose is flat and covered with fine hair. Fur is coarse and thick, in a greyish-brown. Limbs are short and broad, with strong claws.

### Habitat
Leads a solitary life, though tolerates other wombats in a communal situation. Found in forests as well as lightly wooded plains, where the soil type is suitable for burrowing. Digs burrows whose length varies according to conditions; larger burrows measure up to 30 metres. Each burrow can be a network of several chambers.

### Breeding
Breeds throughout the year, though more often in spring. Usually gives birth to a single young. Unfurred, it crawls from the birth canal into a backward opening pouch, where it remains for approximately six months. Out of the pouch, it stays at heel for a further 11 months.

### Food
A vegetarian. Spends most of the night grazing on native grasses, and eating roots of shrubs and trees. Derives moisture from its diet, and can last without water for months.

### Emergency food
Adult: greens such as lettuce; root vegetables, such as carrots and new potatoes; fruit; hay; bran; and stock pellets, such as horse pellets. Short-term feeding: one teaspoon of glucose powder to a cup of water is the best emergency food for a sick wombat.
Baby: either evaporated milk diluted with warm water (ratio 1:1), or one scoop Digestelact (milk powder from a chemist) to 100 ml warm water. Strengthen the formula as the animal gets older. For long-term feeding, add two drops of baby vitamin drops and a pinch of glucose powder to the formula. See FEEDING BABY MAMMALS page 20.

### Handling
Wombats can utter growls and bite, if provoked. Unless you're experienced, all handling should be done wearing heavy-duty gloves. It is better still if the animal can be encouraged to enter the container, placed on the ground in its path. Smaller wombats can be handled through a towel or a cloth. Place over the animal, covering the head. This calms and disorientates it. Scoop up with the cloth and place into the waiting container. A recognised method of handling is by picking up by the torso. Place your arms under the wombat's chest, lifting it with its feet away from you. Always have a transport/ holding container ready, prior to capture. Wash your hands after handling animals.

### Transport

Baby and young marsupials must be kept in improvised pouches for warmth and security. A woollen jumper sewn to make a sack, a small container such as a basket lined with soft cloth, a sheepskin slipper, a woollen hat, are all suitable. Put a lining of soft material between the wool and the animal. Being strong, adults must be transported in a sturdy container such as a heavy wire cage, a metal bin or a large, strong bag. A sick or injured animal may need immobilising and warmth, and may need to be wrapped in a blanket or a towel and nursed while transporting. If transporting in a cage, cover the cage with a cloth to create darkness. That calms the animal and makes it feel secure.

### How to deter

It is rare for wombats to be encountered in an urban environment. In a rural environment it is not the wombats, but their burrows, which can be a problem to stock and property. Stock have broken legs by falling into ground weakened by the burrows. Fences and foundations have also been undermined by the excavations. Unfortunately, not much can be done to deter wombats, as they bulldoze their way into properties. A solid fence or wire mesh dug at least 50 cm into the soil around the property, may be of some deterrence to a burrowing wombat. However, your State Fauna Authority may be of more assistance on the problem. See EXTRA HELP page 193.

### How to attract

Wombats are not commonly found in urban areas, and cannot be attracted to enter a yard. In a rural area, with good burrowing soil covered in native grass, wombats are more likely to appear. Their presence can cause problems, such as ground collapsing above the burrows. This can injure stock and destroy fences and foundations.

You can offer them protection by locking up your pets for the night, as pets are potentially dangerous to nocturnal, native animals.

### Did you know?

*Nature provided wombats with a pouch which opens at the rear, unlike the top-opening one in kangaroos and some other marsupials. One theory is that it is to stop dirt from getting into the pouch when the animal burrows. Or was the wombat facing the wrong way when the pouches were handed out?*

# BANDICOOTS

## Long-nosed Bandicoot *Perameles nasuta*

Bob Miller/NPIAW

Native. Marsupial. Nocturnal.

### Identification

Adult head and body length 31 to 43 cm, with a mouse-like tail. Resembles a rat, with larger ears and a longer nose. General body colour, a greyish-brown above, with a cream belly. Muzzle and ears are long and pointed.

**Habitat**
Leads a solitary life. Found in both tropical rainforests and dry woodland. Rests by day in a shallow nest, which is a scrape in the ground, lined with grass and leaves and covered with debris for concealment. In urban areas its presence may be detected by conical holes about the size of its snout, made in the ground as it forages for food.

**Breeding**
Breeds throughout the year, though more during winter months. Usually gives birth to two to three unfurred young, though occasionally to only one. The young journey into the rear-opening pouch, where they attach themselves to one of eight nipples. They suckle for up to eight weeks, by the ninth week being fully weaned and independent. By such time, another newborn may have begun its cycle of life.

**Food**
Omnivorous. Forages on the ground looking for invertebrates. Also digs holes with its front feet in search of worms, larvae, beetles and other subterranean creatures. Other foods include small lizards, vegetables, berries and roots. Has been known to consume garden vegetables.

**Emergency food**
Adult: insects such as grasshoppers and grubs; worms; dog kibble soaked in warm water; a slurry mixture of bread, pinch of mince, powdered milk and water; finely minced cooked meat; mealworms (available from some bird breeders). Some bandicoots will accept root vegetables such as carrots, and various fruits. Short-term feeding, one teaspoon of glucose powder to a cup of water is the best emergency food for a sick bandicoot.
Baby: either evaporated milk diluted with warm water (ratio 1:1), or one scoop Digestelact (milk powder from a chemist) to 100 ml warm water. Strengthen the formula as the animal gets older. For long-term feeding, add two drops of baby vitamin drops and a pinch of glucose powder to the formula. See FEEDING BABY MAMMALS page 20.

**Handling**
Unless a bandicoot is immobile due to injury, you must be swift to catch one. It can also scratch and bite. Always have a transport/holding container ready, prior to capture. Unless you're experienced, the safest and easiest way to handle a bandicoot is through a towel or a cloth. Throw the cloth over the animal, covering the head. That calms and disorientates it. Scoop up with the cloth and lower into the waiting container, lifting the cloth off. Never handle by the tail as you may injure the animal. You can pick the animal up, wearing gloves. With one hand, grab firmly around the neck and shoulders, and lift, or, place one hand around the neck and the other around the back legs. Wash your hands after handling animals.

**Transport**
Bandicoots can be transported in the same manner as most other marsupials; see COMMON BRUSHTAIL POSSUM, TRANSPORT page 28.

### How to deter bandicoots

In their search for worms and grubs found around roots of plants, bandicoots can dig up parts of a garden. However as they have not adapted well to man's progress, they are not often encountered in suburban backyards. There are no spray deterrents, and the only way to deter them is to 'lock them out'. The yard can be protected by fencing with chicken wire or similar, dug at least 15 cm into the ground, and at least 30 cm above ground. If it is not possible to fence off the entire yard, consider protecting only the vulnerable sections, such as the vegetable beds or flower beds.

If the problem is on a large scale, bandicoots can be trapped and relocated. Trapping is not easy and traps are not readily available. Your State Fauna Authority may be able to help. See EXTRA HELP page 193. Otherwise, you may find somebody reputable who may be interested in trapping the animal for scientific reasons. The biology or zoology departments of a university, or a local zoo are some possibilities.

Should you have access to a mammal trap, you can trap the animal yourself. Bait the trap with earthworms or bread and butter dipped in milk and sugar. Place in the trap and cover the bait with about 5 cm of dirt. Ensure you have obtained a licence from your State Fauna Authority (see EXTRA HELP page 193) prior to trapping. Once trapped, it is your responsibility to release the bandicoot in a suitable area. See RELEASING MAMMALS page 20. Wash your hands after handling animals.

### How to attract

There is no way to attract bandicoots to frequent a yard. They are naturally attracted to areas with plants and shrubs, which house worms and grubs around the root system. Yards with a vegetable patch are more prone to bandicoots than other yards. However, these animals have not adapted to the urban environment as well as other species, and are not frequent visitors to urban yards. If you live on the outskirts of suburbia, have plants in your yard or a vegetable patch, your chances of being frequented by bandicoots are greater. For the bandicoots' protection, lock up your pets for the night, as they are potentially dangerous to nocturnal, native animals.

### Did you know?

*Nature has provided bandicoots with a built-in comb in the form of a grooming toe. Two small toes on the hind foot are joined and clawed, allowing the bandicoot to run them through the hair, combing out any debris and parasites.*

# ECHIDNAS

## Short-beaked Echidna *Tachyglossus aculeatus*

Other names: Spiny Anteater, Native Porcupine

Native. Monotreme. Active by day and night.

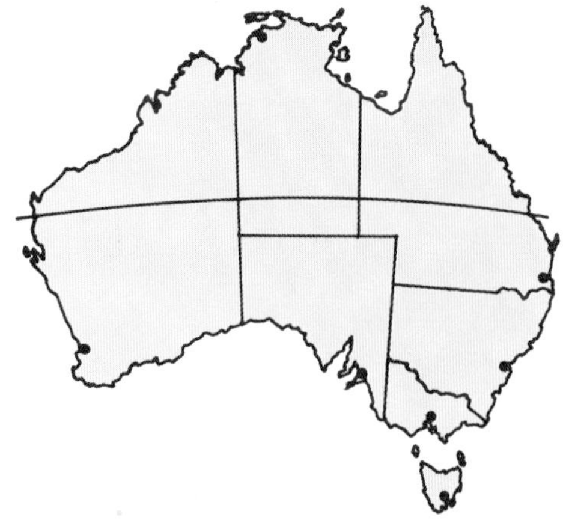

**Identification**
Adult head and body length 30 to 45 cm, with a short tail. Body is furred and covered with light yellow spines with dark tips, to six centimetres in length. Long, naked, tubular snout incorporates a long worm-like sticky tongue. Eyes are small and legs short with long claws.

**Habitat**
Leads a solitary life. Found in all areas wherever there is a supply of ants and termites. Shelters in hollow logs, under scrub, in caves and crevices, and occasionally in burrows.

**Breeding**
Breeds in June to September. Lays a soft-shelled egg into the pouch. The egg hatches after approximately ten days. The furless young remain in the pouch for up to three months, suckling on milk. As echidnas have no nipples, milk is produced by mammary glands on pores of the pouch surface. When the spines begin to grow, the young leaves the pouch and is concealed under cover, where it continues to suckle for the next three months. By six to seven months, the young is independent.

**Food**
Insectivorous. Diet consists mainly of termites and ants. Other insects such as beetles and moths are occasionally eaten. The nests of termites and ants are broken with the front paws, and the ants or termites consumed with the aid of the long, sticky tongue. Parts of nest debris and dirt are also consumed in the process, to be digested and excreted later.

**Emergency food**
Adult: echidnas have a great capacity to store food within their body, therefore can last without food for long periods of time. Do not be alarmed if you cannot feed an echidna for up to 24 hours. As their mouths are so tiny, all food offered must be made to a runny consistency so the echidna can syphon it through the snout. Offer any of the following mixed with water to form a slurry mixture: baby cereal; finely minced beef; hard boiled eggs and evaporated milk; mince meat with glucose powder and bran. Short-term feeding ; one teaspoon of glucose powder to a cup of water is the best emergency food for a sick echidna.
Baby: like the adults, baby echidnas can last without food for up to 24 hours. Due to the shape of their mouths, they are difficult to feed and when orphaned, are tube-fed in captivity. Feeding may be attempted with an eye dropper. Offer any one formula: one scoop Digestelact (milk powder from a chemist) to 100 ml warm water; fresh milk mixed with custard powder to a slurry; or one part powdered milk to four parts warm water with two drops of baby vitamin drops and a beaten raw egg. For prolonged feeding, add two drops of baby vitamin drops and a pinch of glucodin powder to the Digestelact formula.

**Handling**
When disturbed, if the echidna is on a hard surface it will curl up into a ball of spines. Provided you are wearing heavy duty gloves or some cloth protection on your hands, it can be picked up and placed into a container. (Always have a

transport/holding container ready, prior to capturing the animal.) While in that position, it can also be rolled onto a cloth and lifted into the container.

The correct method of handling echidnas is by their hind legs. Wearing gloves if you wish, the echidna can be picked up by the hind legs and held with the spikes away from your body. Echidnas can also be lifted by levering a stick under their body; however care must be taken not to injure the animal's belly. If it is on a soft surface, the echidna may dig itself into the soil. Do not attempt to dig it out with a spade as you may injure it. It is better to wait for it to emerge. Similarly, if it wedges itself under a rock, do not poke at it as you may injure it. Wait for it. Wash your hands after handling animals. For injuries caused by echidnas, see FIRST AID, MAMMALS, page 10.

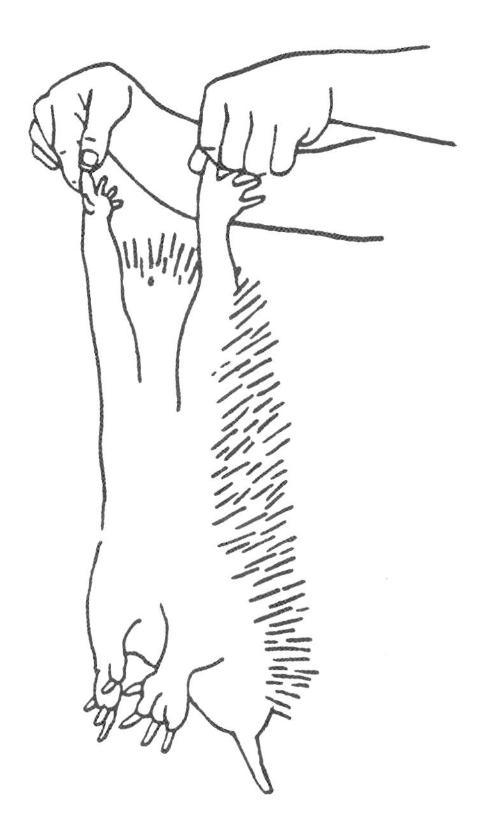

**Transport**

An adult echidna can be transported in a plastic or metal bin provided it is clean, or a stout wooden box. A baby echidna can be transported in a sturdy cardboard box or wrapped in a cloth and nursed. If the young has no spikes, it must be kept warm by wrapping in a soft cloth.

**How to attract**

The presence of this animal is largely dictated by the presence of ants or termites in an area. You cannot consciously attract it to your yard, and unless there is a large ant's nest in the middle of it or you are unlucky enough to be plagued by termites, the only time you'll see an echidna is if it's passing through. Should you suspect that one is frequenting your yard, keep an eye on your pets in the daytime and lock them up for the night if possible, as they are potentially dangerous to native animals.

**Did you know?**

*Australia is home to the only two furred animals in the world that lay eggs: the echidna and the platypus.*

# INSECTIVOROUS BATS

## Gould's Wattled Bat *Chalinolobus gouldii*

D. Ross/NPIAW

Other names: Gould's Lobe-lipped Bat

Native. Nocturnal. Placental mammal.

### Identification

Adult head and body length 65-75 mm. Tail length 4-5 cm. Overall body colour, black to rusty brown, with a lighter belly. A distinguishing feature of this species is a wattle-like lobe on the corners of the mouth.

**Habitat**
Lives in colonies, from a few to about 50 bats. Has a wide distribution throughout Australia, in open forests, and shrubs in suburban areas. Basically a tree dweller, preferring hollows of dead trees. Also roosts in bird nests, under bark and in a variety of man-made structures, such as ceilings, roofs, and basements of dwellings.

**Breeding**
Breeding season September to December, varies regionally. Usually gives birth to twins. Unfurred newborn are attached to the mother's nipples, which are located under the armpit. By one month the young are fully furred.

**Food**
Insectivorous. It hunts in flight, for flying insects such as moths and beetles. It also consumes insects, such as caterpillars, from foliage, as well as ground dwellers such as crickets.

**Emergency food**
Adult: ensure the animal is warm, prior to feeding. Position the bat in an upside-down position, or with its legs lower than the head. Offer a variety of insects such as moths or beetles; baby cereal mixed with warm water to a paste, offered on the end of a brush; or mealworms (available from some bird breeders).
Baby: ensure the animal is warm, prior to feeding. Position the bat in an upside-down position, or with its legs lower than the head. Offer either one scoop Digestelact (milk powder from a chemist) to 100 ml warm water: or evaporated milk diluted with warm water (ratio 1:1). Strengthen the formula as the animal gets older. For long-term feeding, add two drops and a pinch of glucose powder to the formula. See FEEDING BABY MAMMALS page 20.

**Handling**
This bat is capable of scratching when it is being handled. It is therefore advisable to handle it through a cloth, or wearing gloves. Lightweight gloves only, as heavy duty ones may injure the bat. Always have a transport/holding container ready, prior to capture. The safest and easiest way to handle the bat is to place a cloth over its body. Cover the head also, as that will calm and disorientate the bat. Scoop up with the cloth, and gently place into the waiting container. The bat can also be picked up with one hand. Scoop up from the top, enclosing its body within your hand. Ensure that the wings are folded around the body. Wash your hands after handling animals.

**Transport**
See FLYING-FOX, TRANSPORT page 54.

**How to attract**
There is no particular way to attract this species to frequent your yard. Being insectivorous, trees which attract insects may help attract insect-eating wildlife. The best way to help bats and other native animals is to protect their habitat. Plant lots of trees and do not 'tidy' dead branches of trees by chopping them off. Dead branches may contain hollows which make good nesting places for bats and other native animals. Lock up your pets for the night, as they are potentially dangerous to nocturnal native animals.

**How to deter**
If bats live naturally in your area, they may be tempted to settle in the cavities of your roof or walls. The best deterrent is to seal off any possible entry points leading to the cavity. If bats are nesting in the cavity, ensure they have all left the cavity before sealing off. Examine the area in the daytime, and seal off after dark, when the bats have left to forage for food. Another deterrent is to make the area uninviting to bats. The smell of camphor, naphthalene or burning mosquito coils in the cavity area, may deter them. Bats are also not fond of light. Leave a light on for 24-48 hours in the cavity. Ensure that the bulb does not come in contact with any object that may catch fire.

**Bats roosting in your dwelling**
The dark, humidity and quiet, is what attracts bats to roost in roofs, wall cavities and under foundations, particularly in rural areas. Their presence does not cause any danger; however their excreta may stain, and bats can carry mites, though not as harmful as ones found on other animals. If you wish to get rid of bats see HOW TO DETER and seal off all entry points when the bats have vacated for the night. Make a temporary seal first, checking for a few days for noises of any bats left in the cavity. Then seal properly. A colony roosting in a house is not likely to have any offspring there, as those are usually reared in caves.

**Did you know?**
*Some interesting facts about bats: bats are not blind, though some have very small eyes. Bats can swim. There are no blood-sucking, vampire bats in Australia. Those bats are found in Europe.*

# FLYING-FOXES OR FRUIT BATS

## Grey-headed Flying-fox *Pteropus poliocephalus*

Other names: Grey-headed Fruit Bat, Flying-fox

Native. Migratory. Nocturnal. Placental mammal.

**Indentification**
Adult head and body length 23-28 cm. Muzzle is fox-like, ears are large and pointed. Overall body colour is grey, with a reddish fur encircling the neck and shoulders. Legs are furred down to the ankles.

**Habitat**
Lives in camps, formed in summer in varying sizes depending on food supply. Found in forests, mangroves and tropical rainforests, where the bats hang upside-down by the hind feet, on branches of tall trees.

## Breeding

Breeds in October. Usually gives birth to a single, unfurred young. The young is attached to one of two nipples under the mother's armpit, for the first four to five weeks. By such time it is fully furred, and left in the camp at night while the mother forages for food. Upon her return the young suckles. By eight to ten weeks, the young can fly and by three months forages for food independently.

## Food

A vegetarian. Basic diet consists of fruit and flowers of native trees, such as eucalypts, from which it derives nectar and juice. Has been known to consume cultivated fruit, resulting in damage to orchards, particularly when there is a shortage of native food supply.

## Emergency food

Adult: feed the flying-fox while it is in an upside-down position, or while its feet are higher than the head. Offer finely diced soft fruit such as apples, bananas, or oranges, or tinned baby pureed fruit. For prolonged feeding, sprinkle the food with calcium or milk powder. Short-term feeding: one teaspoon of glucose powder to a cup of water, or non-refrigerated pure orange juice pipetted into the mouth with a straw, are best emergency foods for a sick flying-fox.

Baby: feed the flying-fox while it is in an upside-down position, or while its feet are higher than the head. Wrap in a cloth for security. Offer either evaporated milk diluted with warm water (ratio 1:1), or one scoop Digestelact (milk powder from a chemist) to 100 ml warm water. For prolonged feeding, add two drops of baby vitamin drops and a pinch of glucose powder to

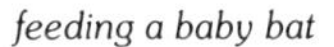

*feeding a baby bat*

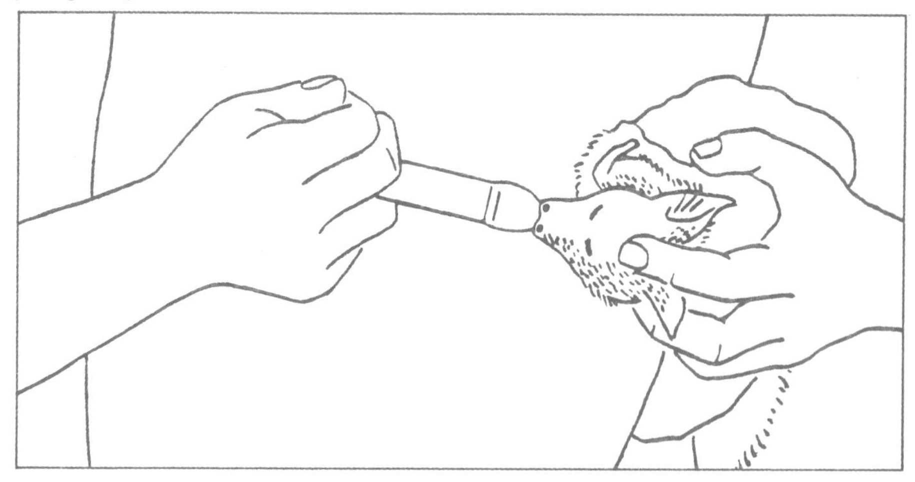

the formula. Strengthen the formula as the animal gets older. See FEEDING BABY MAMMALS page 20.

### Handling

Flying-foxes are capable of biting when handled. Always have a transport/holding container ready prior to capture. Unless you're experienced, the safest and easiest way to handle animals is through a towel or a cloth. Place over the animal, covering the head. This calms and disorientates it. Another method of handling is to grasp the back of its neck with one hand, to immobilise the head and stop the animal biting. With the other hand,

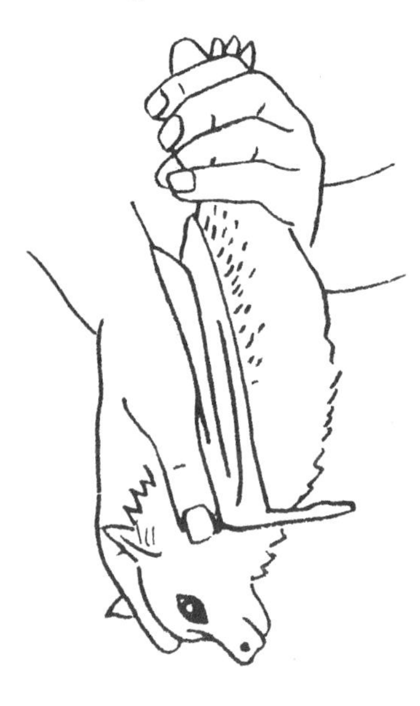

gently unhook the claws, should the bat be hooked. Wrap the wings around the body and gently place in the container. Wash your hands after handling animals.

### Transport

Sick, injured, unfurred or orphaned flying-foxes need to be kept warm. Fold their wings around their body and wrap the animal in a cloth.

*handling a baby bat*

Expose the head and nurse while transporting, or place in a container so that the head is positioned lower than the body. A healthy adult can be transported in a cage or a box. Construct a perch or an attachment inside, from which the animal can hang by the foot, in an upside down position.

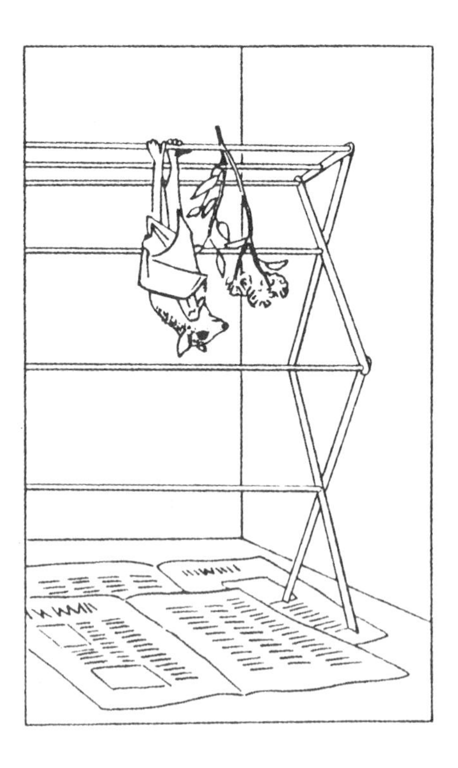

Flying-foxes can also be transported in bags made of strong but soft material. The bag must be big enough to accommodate the animal comfortably. If transporting in a cage, cover with a cloth to create darkness. This calms the animal and makes it feel secure.

### How to deter

Though noise, light and gas deterrents have been used in orchards as means of combating problems with flying-foxes, netting or covering the fruit is still the cheapest and most effective way, for problems on a small scale. A frame of garden netting can be erected over the area, or fruit can be covered with the net, or paper or plastic bags. To prevent a colony from settling on the fruit, the first 'leading' bats, which fly in to 'inspect' the possibilities, should be deterred from settling. If at all possible, pick the fruit before it ripens, to save it from predators. Remove any fallen fruit from the ground, as it entices predators.

Larger establishments, such as orchards, should seek expert advice on the problem. Some major pest control companies are geared to deal with the problem.

### How to attract

You cannot attract flying-foxes to your yard, unless you live in an area where they are found naturally. However, a word of caution. They can be pests and they come as a package deal, a group or nothing. They are noisy, messy and fond of cultivated fruit. They also know no boundaries, and your problems may be passed on to your neighbours. However, this otherwise delightful mammal will benefit from the following trees: native fruits, such as native figs; eucalypt blossoms such as *Eucalyptus robusta*, red bloodwood and any other fruit- and nectar-producing trees. Ensure that all pets are locked up for the night, as they are potentially dangerous to nocturnal native animals.

### Did you know?

*Flying-foxes have a sophisticated system of communication, based on more than 20 different calls. Each call is unique and has a meaning or conveys a message. Some of these calls signify danger, distress, keep away from this area, or come to this area, to name a few.*

# MICE AND RATS

## In the house

Droppings, teeth marks, tracks in the dust, disappearance of food, noise, and odour are some signs of the presence of mice or rats, apart from the obvious sighting of the animal. Apart from being a health risk, their nesting material and their gnawing on electrical wires are fire hazards. Expert advice may be needed to identify the species by the faeces, and for dealing with the problem. See EXTRA HELP page 193. Presence of rodents can be tested by sprinkling talcum powder or flour overnight, on the floor of the suspected area. Check the next morning for tracks. Various rodent baits and traps are available on the market, see TRAPPING MICE AND RATS page 000. Having eliminated the problem, ensure all entry points through which mice or rats are entering are sealed off. Keep in mind that mice and rats are capable of squeezing through tiny holes, and can climb walls, and jump. Keep the house free of food spillages and piled debris, such as newspapers, which provide material for nesting. For large infestations, it may be necessary to engage a pest control company to deal with the problem.

## Trapping

Trapping is the most popular method of dealing with mice and rats, for problems on a smaller scale. Traps are readily available on the market, and when setting, keep the following in mind: the best time to put traps out is after dinner, from 7 pm to 10 pm. As mice and rats dislike crossing open spaces, traps are best positioned near the runways (edges of walls) or under cover, not in the middle of a room. Put out as many traps as you can, covering the whole of the infested area. The more traps, the quicker you'll catch the rodents. Choose a suitable bait, as contrary to common belief, cheese is not a preferred food. Use chocolate, bacon, pumpkin seeds, prunes, peanut butter on bread, fish, nuts, rockmelon, apples, or meat. Use any one at a time or a combination of all. Once the traps have been set, vacate the area and keep the noise down. Return every hour or so to check the traps and remove the bodies. The traps can be discarded with the animal, or the animal taken off the trap and the trap re-used. Once the problem has been eliminated, it is important to seal off any entry point through which the rodents may have entered the house. Hygiene must also be observed, wiping any spilled food and storing food in tight containers.

## How to deter

Search for food and shelter are the main reasons why mice and rats may enter homes. They are more likely to try and fulfil these needs during colder months. As they smell, pollute with their excreta, can transmit diseases, damage clothing, food, books and chew on electrical wires, their presence is most undesirable. Hygiene and tightly constructed structures are the main deterrents. In areas prone to mice or rats, do not store garbage in plastic bins, as rodents can chew through them. Wipe off any food spills immediately, remove all uneaten pet food, and put tight lids on garbage bins. Block off any possible entry points through which mice or rats may enter the house. Ensure all doors are tight fitting, and windows are screened. Examine the guttering, as that is another possible entry point. Mice and rats have a natural dislike for crossing open spaces. That can be utilised to your advantage. Make a concrete border around the house and keep it free of stored material. A trim garden with cut grass, may also deter mice and rats from crossing into the dwelling. Blocking off any possible entry points to the house, is most important after treating the place for mice or rats.

# INTRODUCED MICE

## House Mouse *Mus musculus*

R. Whitford/NPIAW

Other names: Mouse

Introduced. Nocturnal. Rodent.

### Identification

Adult head and body length 6-9.5 cm. Tail length equal to body length. Muzzle is pointed, eyes small and ears large and furry. Body is covered in fine fur, a shade of grey above, with a lighter belly. Introduced species can be distinguished from native species by the presence of a notch on the inside back of the upper teeth, into which the lower teeth fit. The House Mouse has three pairs of teats, while the native mouse has only two pairs.

**Habitat**
Lives in small groups. Has adapted well to the environment, and can be found throughout Australia. It is most commonly found in areas where there is a food supply, such as storage barns, restaurants, hotels, and field-crops. It shelters in burrows in the open, and builds nests in human habitation areas. Nests can be made of shredded paper and are usually placed under debris, between logs, in straw or in foodstuffs. Mice are more likely to enter dwellings in cold weather, and prefer a dry environment.

**Breeding**
An opportunistic breeder, taking advantage of most favourable times. Young are born unfurred, averaging four to eight per litter, though up to 12 are not uncommon. By 18 days they are fully weaned, and independent by 21 days. Females can have up to 12 litters per year, which can result in up to 1000 offspring per year from one pair of mice.

**Food**
Omnivorous and an opportunistic feeder. Eats a variety of food, showing a preference for cereal grain. Also feeds on insects, fungi, moss, meat, fish and vegetables. Moisture is derived from food, and the mouse needs very little, or no water. Being a rodent, it has a need to gnaw on hard materials to wear down the incisors, in the process frequently causing damage in a commercial or domestic situation. Mice are also responsible for damage to stored food stuffs.

**Handling**
Mice can transmit diseases such as salmonella poisoning and ringworm to man and pets. Therefore, they should not be handled with bare hands, whether the animal is dead or alive. Handle with gloves on, or through a tissue or cloth. Dead mice can be picked up by the tail, or scooped onto a dustpan with the brush or some implement, and disposed of as soon as possible. Live mice can be lifted with one hand, by the back of the neck. That immobilises the head and stops the mouse from biting. Wash your hands after handling animals.

**Did you know?**
*In terms of adaptation and distribution, mice are the world's most successful mammals, apart from man. Their ability to live and breed in the hottest and coldest temperatures is undoubtedly a contributing factor. Mice were probably introduced into Australia around the time of the First Fleet, aboard ships.*

# Introduced Rats

## Black Rat *Rattus rattus*

K. Stepnell/NPIAW

Other names: Roof Rat, Tree Rat, Ship Rat, European Black Rat, Fruit Rat, White-bellied Rat

Introduced. Nocturnal. Rodent.

### Identification

Adult head and body length 16.5 to 20.5 mm. Tail length 18-24 cm. Tail is longer than the body. Untrue to name, there are very few rats which are jet black. Body colour varies from black to shades of light brown and grey-brown.

### Habitat

Lives in colonies. Has adapted well to the environment and can be found along the entire coast of Australia. Greatest numbers are found near human habitation, where it lives in rubbish dumps, old buildings, around farms and in warehouses. It must have daily access to water, and lives near waterways or within reach of water, whether in urban, rural or native environment. It shelters in nests made of paper, rags, or grasses, built usually above ground in dense vegetation, scrub or buildings. It is more likely to enter buildings in cold weather.

### Breeding

An opportunistic breeder, taking advantage of most favourable times. Young are born unfurred, averaging five to ten per litter. Six litters per year is the average, though litters can vary from three to 12 in any year. By 20 days, the young are fully weaned and independent.

### Food

Omnivorous and an opportunistic feeder. Eats a variety of food, showing a preference for fresh fruit and vegetables. Daily diet must contain water, and if not available at the nesting place, the rat will travel long distances to quench thirst. Being a rodent, it has a need to gnaw on hard materials to wear down its incisors. That frequently results in damage to domestic and commercial properties. Rats are also responsible for damage to stored food stuffs.

### Handling

Rats can transmit diseases such as salmonella poisoning and ringworm to man and pets. Therefore they should not be handled with bare hands, whether the animal is dead or alive. Handle with gloves on, or through a tissue or cloth. Dead rats can be picked up by the tail, or scooped onto a dustpan with the brush or some implement, and disposed of as soon as possible. Live rats can be grabbed with one well-protected hand, by the back of the neck. That immobilises the head and stops the rat from biting. Wash your hands after handling animals. As rats can carry diseases, medical advice and a tetanus needle are advisable. If bitten. See FIRST AID, MAMMALS page 10.

### Did you know?

*Fleas on rats were responsible for the loss of millions of lives throughout history. In the early 1600s, more than 150 000 lives were lost in London alone. Rats, and man's lack of knowledge on the importance of hygiene, contributed to the spread of a germ called* Yersisia pestis, *which resulted in the Bubonic Plague.*

# BIRDS

Birds contribute to the elimination of insect pests and some, such as the honeyeaters, are useful pollinators. Next time you see a bird, take a good look at it. Its beak is a good indicator of what type of food it eats. Strong, hooked bills, as in birds of prey, are used for tearing the flesh of animals. Long, thin bills, as seen on the honeyeaters, are good for extracting nectar from tube-like flowers. Small, wide bills are efficient for catching insects in mid-air, as swallows do. Wide, flat bills, like those of ducks, are useful for skimming the water surface in search of plant or animal matter, and strong, hinged bills will crush seeds and nuts, as do parrots. Birds' feathers also serve many purposes: as well as being necessary for flight, they insulate against the heat and cold, they provide waterproofing, act as a camouflage, are useful in displays of courtship and aggression, and act as a soft lining for a nest.

And how do birds fly? It's a very complicated and not fully understood procedure, but underneath the feathers is a light skeleton with hollow bones or bones made of a honeycomb structure to make the bird light enough for flight.

## How to deter

If birds become a problem, you may wish to deter them from entering your yard. If you have no trees on your property, your chances of being faced with bird problems are smaller, although some urban species of birds may still be attracted to your yard or your dwelling. Depending on type of bird, consider the following suggestions:

Eliminate possible roosting places by treating roosting areas such as window ledges. There are sticky gels on the market, used for coating the ledges in order to make roosting uncomfortable. Prevent birds from nesting in your roof by eliminating any possible entry points, sealing all cracks and holes leading to the roof area.

Hang strips of tin foil or coloured paper in the area where birds are a problem. As they flap in the breeze, they may deter the birds from

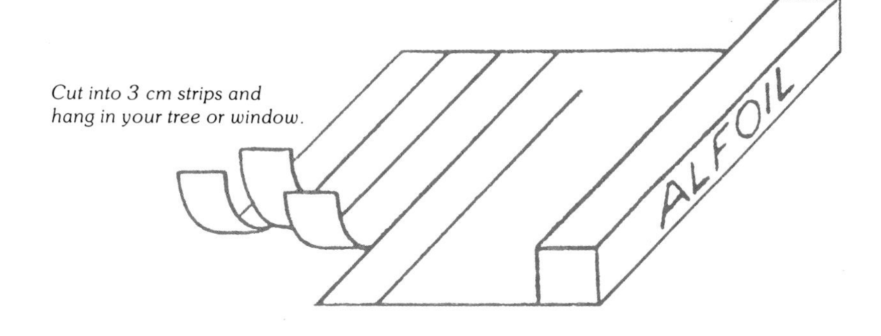


*Cut into 3 cm strips and hang in your tree or window.*

approaching. Where possible, put bags or netting over fruit to let it ripen without bird interference. Cover fishponds with chicken mesh or netting. Using cardboard, plastic or masonite, make a silhouette of a bird of prey, such as a hawk. Paint it black or roughly to resemble the bird and hang to flap in the breeze. Although cats will certainly deter birds, they are potentially dangerous to most birds, particularly nestlings.

If all else fails, contact a reliable pest control company. They may be able to advise you or treat the

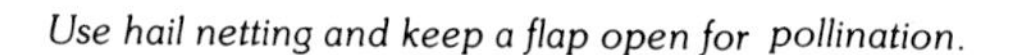

*Use hail netting and keep a flap open for pollination.*

*Make model the same size as bird i.e. about ½-1 m wide.*

problem, for a fee. Some major pest control companies sell and hire noise machines to deal with bird problems on a larger scale. For major problems such as birds attacking crops or orchards, expert advice on the subject should be sought. See EXTRA HELP, page 193.

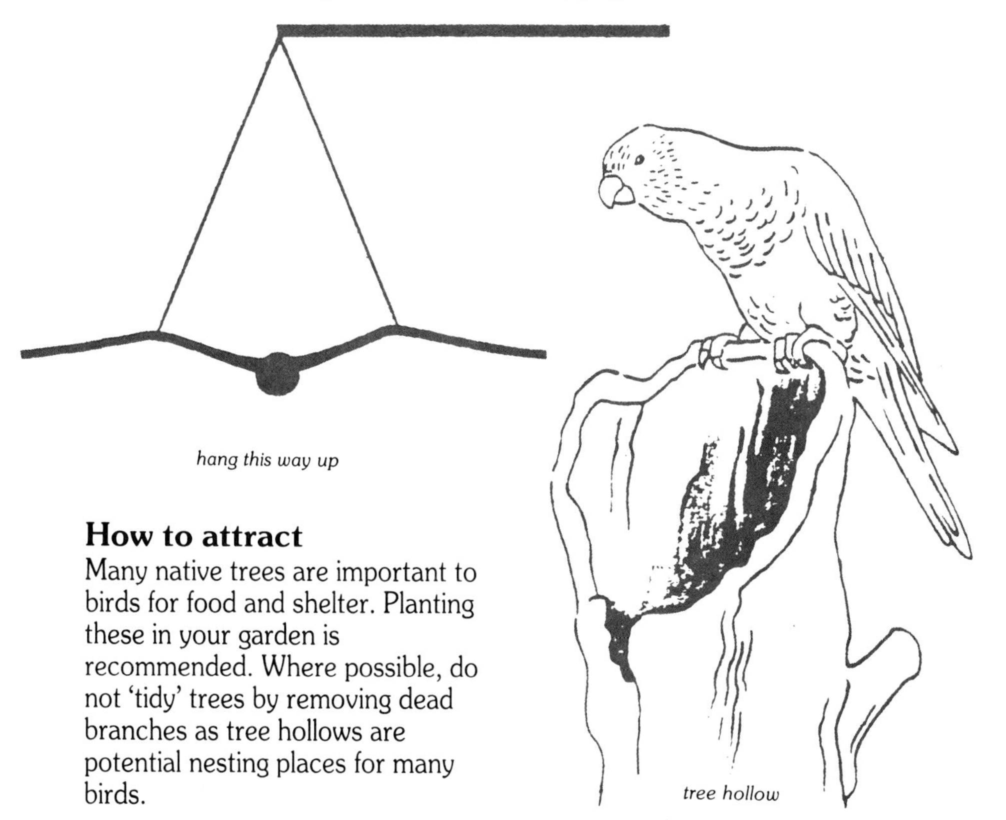

*hang this way up*

*tree hollow*

## How to attract

Many native trees are important to birds for food and shelter. Planting these in your garden is recommended. Where possible, do not 'tidy' trees by removing dead branches as tree hollows are potential nesting places for many birds.

Erect a bird bath, preferably near vegetation. An inverted garbage-can lid or a large pot-plant saucer are suitable. Mount at least a metre off the ground or in a way as to stop cats' access.

Some species of birds will happily nest in a yard, if provided with a nest box. Contact your local Bird Observers Club or an equivalent organisation for information on which birds accept nest boxes, and types of boxes to erect. A word of caution however, it is not uncommon for other types of birds, including some parasitic species, to occupy a nest box.

Do not keep cats. If you have a cat, put a collar with a bell around its neck to act as a warning to wildlife. Lock the cat up for the night if possible, as it is potentially dangerous to wildlife.

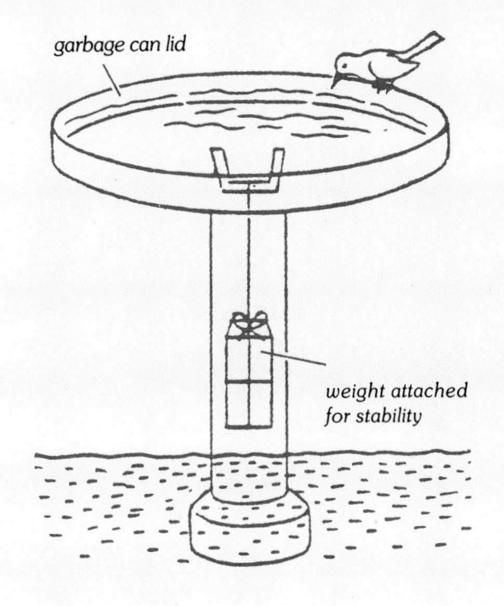


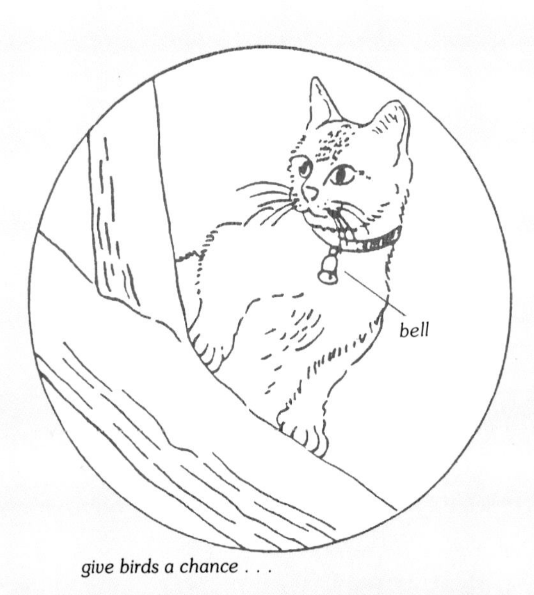


*give birds a chance . . .*

## Handling

Most birds are safe to handle, provided you stay clear of the bill and claws. Like all animals on the defensive, they will attempt to bite, and their claws can be reasonably sharp. If injured they may be more aggressive, though some injuries may make them less mobile. Always have a holding/transport container ready prior to capture. Unless you're experienced, the safest and easiest way to handle birds is through a towel or cloth. Place over the bird's body, covering the head. This disorientates and calms the bird. Placing your hands on both sides of its body, scoop it up with the cloth. Should the claws be uncovered, face them away from your body. Have the transport/holding container ready. Gently lower the bird inside, lifting the cloth off as you do so.

When handling larger birds with bare hands, place one hand around the bird's body, immobilising the wings and feet. With the other hand, hold the back of the bird's neck to immobilise the head. Lift the bird, taking most of the weight in the hand holding the torso.

When handling smaller birds with bare hands, the bird can be taken with one hand so as to accommodate the wings, claws and tail. When held firmly, this method restricts movements of the head and shoulders. Hold away from your face.

Gardening gloves can be worn for extra safety when handling larger birds. Lightweight gloves only, or bare hands, for smaller birds. You risk injuring the bird if you use heavyweight gloves when handling smaller birds. See individual entries on birds for more information.

Wash hands after handling animals.

## Transport

Birds may be transported in a cardboard box big enough to accommodate the bird comfortably. **Punch holes in the lid or at the side for airflow**. If the lid is not tight-fitting, or if the box does not have a lid, cover securely with a flat object such as a book, to keep the bird from escaping. Do not line the box with any material, as the bird may become entangled—unless, that is, the bird has been injured and must remain wrapped up.

Bigger birds, with stronger claws and beaks, may need to be transported in a cage or, in the short-term, in a metal garbage bin. This is necessary for efficient chewers like cockatoos and galahs. Be sure that the lid of the garbage bin is secure, otherwise the bird will escape. But make sure that the lid does not completely cover the bin opening, or the bird will suffocate.

A large cardboard box with a wax coating is another short-term solution for bigger birds; punch holes for air.

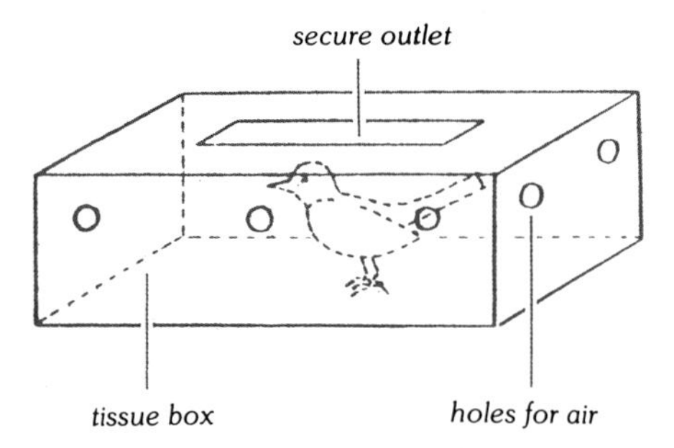


For tiny birds, an empty ice-cream container is another option; once again, do not line the container, and do punch holes for air. See individual entries for information specific to particular birds.

When transporting birds in a cage, cover the cage with a cloth to create darkness. This calms the bird and makes it feel more secure.

## Birds flying into windows

A bird may fly into a window for two reasons. It may see an outlet on the opposite side and think it can fly through, or it may see its reflection in the glass, assume that it is an intruder and attack it. Depending on the impact, the bird might die; it might suffer body injuries, or shock; or it could fly off, uninjured. If the impact has not killed the bird, but it was unable to fly off, try to capture it. If there are obvious body injuries, take it to a vet. If there are no obvious injuries, treat the bird for shock. TREATMENT FOR SHOCK page 67. If the bird has not improved after approximately one hour or if its condition deteriorates, take it to a vet.

In order to prevent birds from flying into windows, you must remove the 'mirror' from the window or deter the bird from approaching it. Draw the inside curtain or hang an awning or external blind over the window. Coat the window with a cream cleanser, such as Ajax or Bon Ami. It will harden. Leave on the window for a few days or till the bird has left the area. Using cardboard, plastic or masonite, make a silhouette of a bird of prey such as a hawk. Paint it black or roughly to resemble the bird and hang it outside the window to flap in the breeze (see page 62-63). Hang strips of coloured paper or tin foil outside the window; their flapping in the breeze may deter the birds from approaching. Erect a screen of chicken mesh or garden netting in front of the window. Temporarily, stand something in front of the window, such as a trellis, or hang a bedspread or blanket over the window. **Note:** The problem of birds flying into windows is highlighted with the use of shiny sun-protection film over the window.

## Symptoms of sick birds

Unless you are particularly experienced in the handling of birds, it may be difficult to detect that a bird is sick. A lot of birds are victims of the 'great white hunter' syndrome, where well birds are 'rescued' by people who decide they are ailing. Young or immature birds are the most common victims, as often they appear sick, abandoned or helpless as they wait for their parents which are out foraging for food. If the bird is close enough to observe, any of the following symptoms may indicate that it is sick: droopy posture, feathers fluffed up, or a tilted head. If the bird is on a branch and you are uncertain of its condition, a gentle squirt with the hose will soon make a well bird fly away. A sick bird is more likely to remain, or move up a little on the branch. If possible, capture the bird and take it to a vet for examination.

## Internal injuries

It is not always obvious that a bird has internal injuries. Birds that fly into a window pane, are hit by a car or grabbed by a cat, are quite likely to have suffered internal injuries. If any of the following symptoms occur, the bird may have internal injuries: if the bird is gasping for air, or there is a noise such as rasping or rattling during the breathing. If uncertain, it is better to take the bird to a vet for examination. You may wish to settle the bird first by treating for shock. See TREATMENT FOR SHOCK.

## Treatment for shock

Treatment for shock is used to pacify the bird or settle it down. The procedure can be applied to almost all injuries, or when the bird is distressed, prior to examining or treating it. It is based on three objectives: noise reduction, maintaining warmth and light reduction. Find a container big

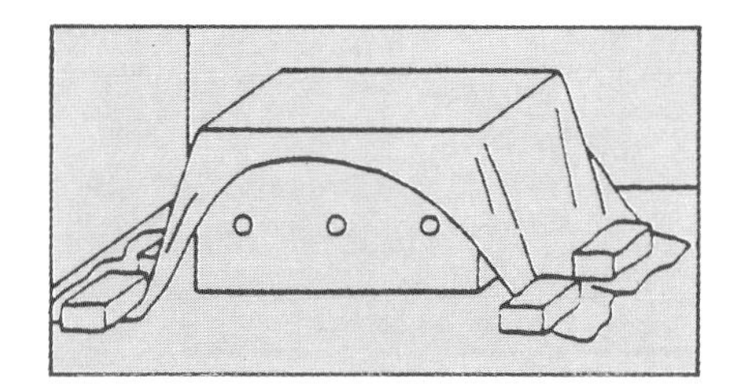

enough to accommodate the bird comfortably. Line the bottom with cloth for added warmth, ensuring the bird cannot become entangled. If airtight, punch holes at the sides or in the lid to ensure adequate air-flow. Place the bird inside the box and cover with the lid or a cloth. If using a cloth, secure so it will not lift. That stops the bird from escaping and eliminates light.

Place the box in a quiet spot away from excessive noise or vibration. On a cold day, place in a warm room if possible. Leave the bird alone for at least an hour. Examine or disturb as little as possible, and only when the bird appears to have settled.

Having seen some improvement in the bird, provide a source of energy. The most readily available one is a mixture of water and glucose, honey or brown sugar. Mix at a ratio of 9:1 for larger birds such as galahs, or 5:1 for smaller birds such as pardalotes. Offer the liquid in a container. Dip the bird's beak in and wait for it to drink. Never pour liquid down the bird's throat, as it may choke. When treating for shock, use your commonsense to create a suitable environment. A box may not always be available, or the treatment may need to be carried out while travelling in a car. Always keep in mind the three necessities: minimal noise and light, and added warmth. Wash your hands after handling birds.

## Birds hit by a moving vehicle

A bird may fly at a car accidentally or in pursuit of an insect lit up by the car's headlights. If the bird hits the car, stop the car and try to find the bird. Park so as not to inconvenience other motorists, and ensure you do not risk being run over in the process. If the bird is obviously dead (chest not moving, staring eyes, does not respond to touch), remove it from the road. Consider taking it with you to donate to a museum, see DEAD BIRDS page 75.

If the bird is not dead, it is your moral duty to help it. You will need to handle the bird, and that is best done through a towel or a cloth, or any piece of old clothing. Place over the bird, covering the head. That disorientates and calms it. See HANDLING page 64. Scoop the bird up with the cloth, and place into a box or nurse it, if no box is available. (When travelling long distances on country roads, it is sensible to carry a cardboard box and an old towel in the boot of the car. These items are useful in capturing and transporting injured wildlife). Ideally, the bird should be taken to the nearest vet for examination. However, if no vet is available or there appear to be no major injuries, treat the bird for shock. See TREATMENT FOR SHOCK page 67. Keep in mind that the bird may have suffered internal injuries (refer INTERNAL INJURIES page 67). Should the bird suffer irreparable damage, euthanasia may be the most humane solution. Take the bird to the nearest vet, who will put it down after confirming that the bird cannot be saved. In the country, far from any help, it may be necessary for you to perform the task. See EUTHANASIA page 75. When you rescue a bird that has been injured by a car, keep in mind that certain birds should ideally be returned, after recovery, to the same area. Territorial birds such as kookaburras may kill other kookaburras released into their territory (refer KOOKABURRAS, DID YOU KNOW? page 77). It may be helpful to take a note of the area where the bird was rescued. Wash your hands after handling birds.

## Body injuries/bleeding

Birds do not have a large blood volume and do not have to lose a lot of blood before a serious problem develops. If you have witnessed the accident, use your judgement as to whether there is a possibility of internal injuries. Small wounds are best left untreated as handling could be more traumatic to the bird than the wound itself. Unless they injure a foot, superficial wounds generally do not cause birds much pain. If you have captured the bird, treat for shock. See TREATMENT FOR SHOCK page 67.

Some superficial wounds can be treated by the layperson. Mix a solution of ½ teaspoon of salt to a cup of water and dab onto the wound with a cotton ball. For serious wounds, or if the bird's condition deteriorates, take to a vet. Wash your hands after handling birds.

## Leg injuries

Leg injuries in birds cause them pain, and hinder their search for food and their escape from predators. Treatment should be left to an expert. Try to catch the bird.

See HANDLING page 64. If the bird is distressed, or a vet is not available immediately, treat for shock. See TREATMENT FOR SHOCK page 67.

## Wing damage

A bird with an injured wing risks being caught by predators and cannot hunt for food efficiently. Try to capture the bird. (See HANDLING page 64). Take the bird to a vet for examination. The bird will arrive in much better condition if you strap it first. This is done by positioning the wing back in place and strapping the bird's torso with a cloth, bandage or masking tape. Wrap firmly as you would for a sprain in humans. Offer the bird a drink. Mix water and glucose, honey or brown sugar in a ration 9:1 for larger birds such as galahs, and 5:1 for smaller birds such as pardalotes. Offer the drink in a container. Dip the bird's beak in and wait for it to drink. Never pour liquid down the throat, as the bird may choke. Lie the bird on its side in a box big enough to accommodate it comfortably, and take to a vet within 24 hours. If the wing is broken, it can be repaired by many methods, such as pinning through the hollow bones. Wash your hands after handling animals.

## Waterlogged birds

After heavy rain, many birds take on an 'unhealthy' appearance. They may sit on the ground instead of perching, do not fly away when approached, or generally have that washed-out look. Some are covered in mud. Capture the bird if possible, see HANDLING page 64. If the bird flies away when approached, then you may well

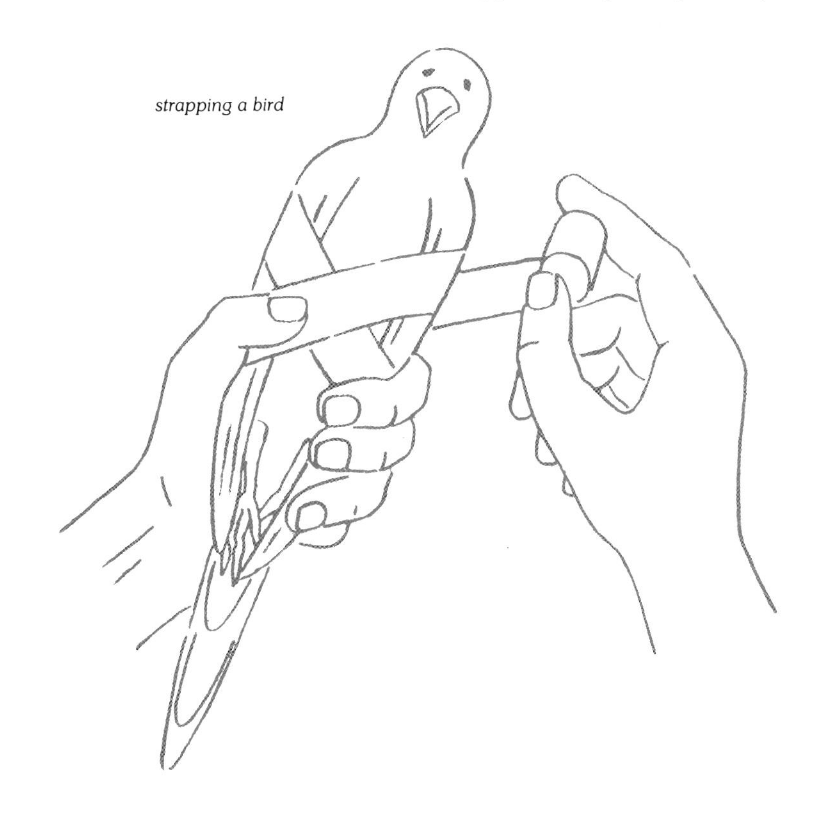
*strapping a bird*

have overestimated the severity of the problem. If the bird is covered in mud, sponge it clean with warm water and dish detergent (not dishwasher), diluting one dessertspoon of detergent to a litre of water. Dry the bird thoroughly. Warmth is important. Place the bird in a warm, quiet place; see TREATMENT FOR SHOCK page 67. Having pacified it, offer it food (see EMERGENCY FOOD on pages relevant to the species). Or you can offer it a source of energy, see TREATMENT FOR SHOCK page 67. for the formula and the procedure. As birds found in this condition may have contracted pneumonia, it is advisable to take the bird to a vet for examination prior to release. Wash your hands after handling birds.

## Birds covered in oil

Birds, particularly those whose habitats are near watered areas, often perish when their feathers become clogged with oil from an oil slick. When this happens in large numbers in a particular area, organised rescue programmes are necessary. Some birds are so severely affected that euthanasia is the most humane treatment. Should you find a bird covered in oil, it must be cleaned as soon as possible. If a delay is inevitable, the bird must be stopped from preening itself, as in the process it comes in more contact with the poisonous substance. The bird may be prevented from preening by putting a collar around its neck. Using cardboard or light plastic (such as an empty margarine container), cut a circle with a circular cutout. Size will depend on the bird, the circumference of the inside circle being roughly the size of the bird's neck. Place around the bird's neck and staple or fasten with a tape. Mix a solution of one dessertspoon of dish detergent (not dishwasher) to a litre of warm water, and bathe the bird repeatedly till it is free of oil. In heavy contaminations, cornflour sprinkled on the bird prior to washing may help draw out some of the oil. Dry the feathers gently but thoroughly. Hold the bird for at least a day, until it has accepted some food (see EMERGENCY FOOD on page relevant to the bird species), preened itself and dried. If at any time during the contamination the bird has swallowed some oil, it may be sick. Observe it and take to a vet if it does not improve. Wash your hands after handling birds.

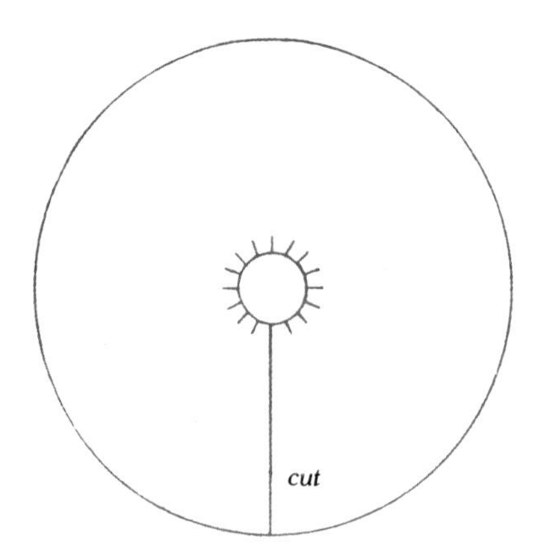


*a preening collar*

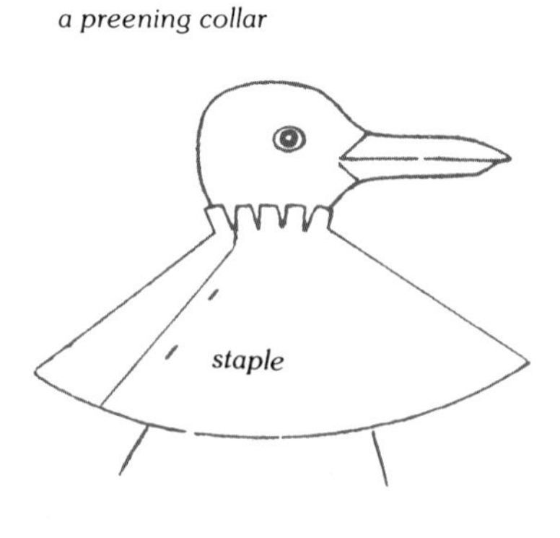

## Feathered birds out of nest

A young, feathered bird out of the nest is not necessarily in trouble. Many fledglings leave the nest well before they are able to fly or fend for themselves. Though it may not always be evident, the parents are never far away, always ready to feed, protect and guide the young. As a result of the 'great white hunter' syndrome healthy young are often separated from their parents. If you see young feathered birds wandering, the first thing to do is to lock up your pets. Before declaring the birds abandoned, observe them for a few hours from a distance, giving the parents a chance to appear. Stay out of sight, as the parents may be waiting for you to leave before approaching the young. One sign of the presence of parents is an exchange of chirping between them and the young.

If you can see the nest and it is within your reach, place the young back in the nest, handling them as little as possible. Otherwise, construct a fake nest as per instructions in UNFEATHERED BIRDS OUT OF NEST. Place the birds inside the fake nest and hang in the area where the young were found or near the original nest. It is possible for the parents to feed the young in the fake nest. If the young are at the age of leaving the nest, they will soon leave the fake nest too. At that point, they are best left alone for nature to take its course. Keep your pets away from the young. If the young have been truly abandoned, see EXTRA HELP page 193, for help re fostering them. It is also possible for the mother to evict the young from the nest, if she senses that it is sick. You may wish to take the abandoned bird to a vet for examination, or foster it, observing its condition. Wash your hands after handling birds.

## Unfeathered birds out of nest

If an unfeathered bird is out of its nest, it has either fallen out accidentally, or has been evicted because it is ill. As there is no way of knowing which is the case, the aim is to reunite the bird with the mother or foster it. If the nest is within your reach, gently place the bird inside with minimal handling.

If the nest is too high, make a fake nest to accommodate the bird.

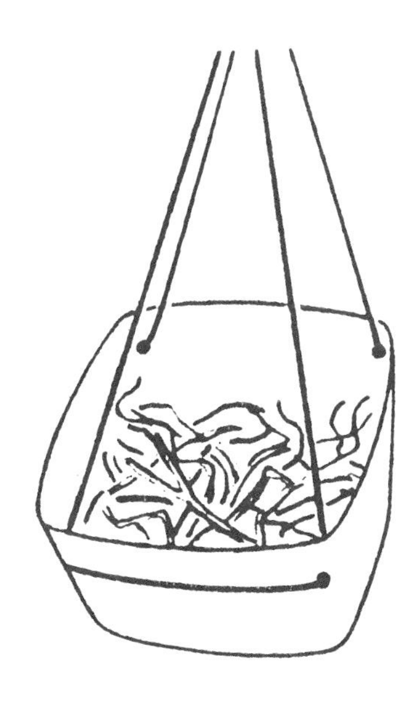

Before proceeding with the fake nest, place the bird in a warm, quiet place away from pets. An empty ice-cream container makes an ideal fake nest. Punch about ten holes at the bottom for drainage. Line the container with grass to create comfort and enough depth for security. Punch a hole in each corner and thread string through to hang the nest. Gently place the bird/s inside. Hang the nest on a branch in the area of the original nest or where the parents are hovering. Do not hang near a forked branch as cats will have access to it. Observe the nest from a distance giving the parents a chance to find it.

If after a day the parents have not shown, treat the bird/s as abandoned. Take the nest down and feed the young. See FEEDING YOUNG BIRDS, and EMERGENCY FOOD on page relevant to the bird species. Call for help regarding fostering the young. See EXTRA HELP page 193.

In order to feed the bird the correct food, it must be positively identified. When unfeathered, it is not an easy task. The safest way to identify the young is to identify the parents. Observe the birds in the area or refer to descriptions of the young in this or another bird book. Should you foster the bird yourself, guidance from more experienced people will be needed. Unfeathered, abandoned birds have a high mortality rate, so do not feel a failure should it die in your care or in the care of someone experienced in fostering birds. Wash your hands after handling birds.

## Feeding young birds

Feeding young birds is a demanding job. In a natural situation, the parents may frequent the nest with some food every ten minutes. Some young, such as baby magpies, show their readiness to accept the food by tilting their heads up and gaping their beaks. When being fed in captivity, they

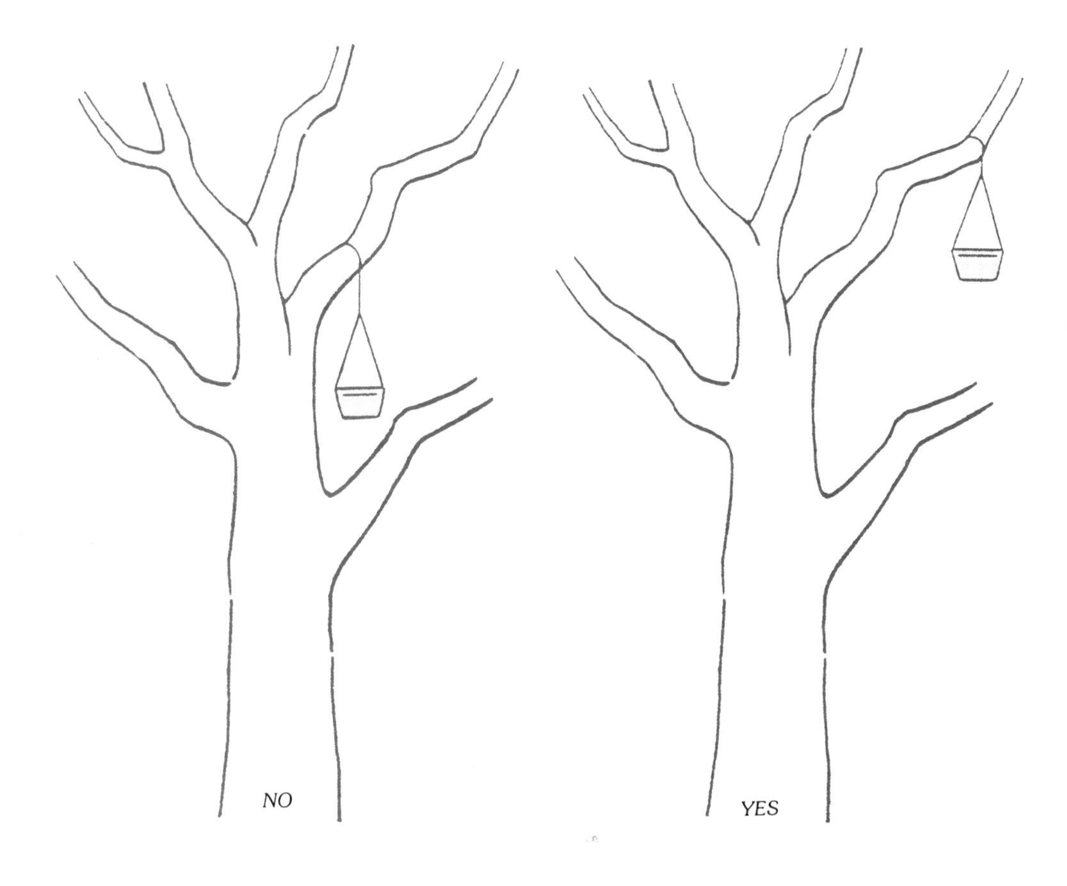

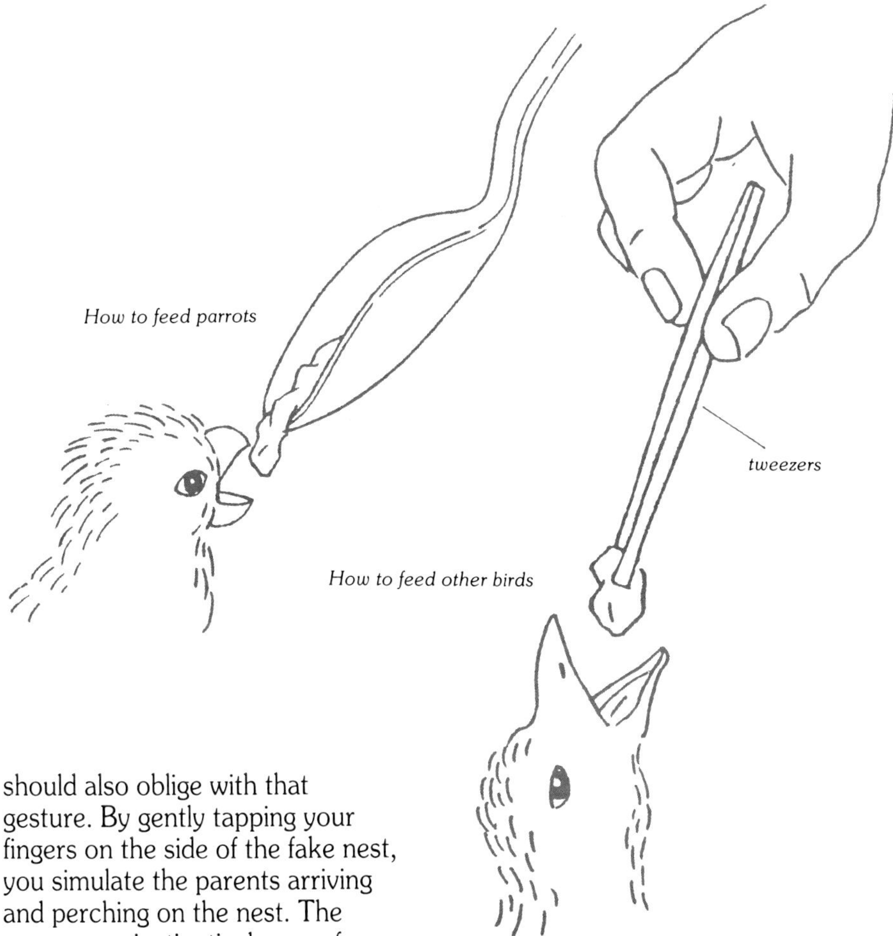


should also oblige with that gesture. By gently tapping your fingers on the side of the fake nest, you simulate the parents arriving and perching on the nest. The young may instinctively gape for food. Small quantities of solid foods can then be dropped into the beak, using tweezers, a toothpick, a teaspoon with bent edges, or your fingers, having first washed your hands.

Some young, such as those from the parrot family, do not gape for food. Food in a slurry consistency can be offered to these birds on a teaspoon. Gently touch the bird's beak with the feeding spoon or tap your fingers on the side of the fake nest, to simulate mother's arrival and get the bird to accept the food off the spoon.

Very young, unfeathered birds may need to be crop-fed using a syringe with rubber tubing. It is a specialised procedure and should be left to experienced people. Should you wish to tackle the job yourself, advice will be needed from experienced people. See EXTRA HELP page 193

When feeding young birds, food such as tinned pet food should be ground to a consistency suitable for dropping into the beak, or mixed into a slurry, such as with baby cereal mixed with water, to be offered at the end of a teaspoon. Water should be made available only as part of the food mixture and never poured down the bird's throat as the bird may choke.

Keeping the bird clean from food is most important. Do not let food build up around the beak or anywhere on the feathers as bacteria can breed in the built-up food resulting in the bird becoming sick. Keep the feeding utensils clean, preferably sterile. Sterilising fluids used for infants' bottles are suitable.

Some young birds have to be force-fed at first to establish a new routine. Check with a vet first, to ensure the bird is not refusing the food due to sickness. If the bird is eating well and suddenly loses its appetite, do not force-feed. Take it to a vet as it is a sign that it may not be well. As the bird grows, encourage it to help itself to the food. Start by leaving the feeding utensils with some food on them, within the bird's reach. For long-term feeding or bird rearing, advice from experienced people will be needed. See EXTRA HELP page 193 Wash your hands after handling birds.

## Pesticide poisoning

Birds, particularly insect eaters, can be the victims of pesticide poisoning. It is easier to detect when a group of birds in one area suffer from the same symptoms. It is more difficult to detect in a single bird. Some more obvious signs of poisoning are: wobbling or loss of limb control, listlessness, in more severe cases convulsions or tilting back of the head. Initially, treat for shock—see TREATMENT FOR SHOCK page 67. Take the bird to a vet. If you suspect malicious poisoning, or if many birds have been poisoned in one area, notify your State Fauna Authority. See EXTRA HELP page 193. Wash your hands after handling birds.

## Releasing birds after captivity

Releasing a bird is not just letting it go, but releasing it into a suitable environment. Prior to releasing, learn about the bird, its natural habitat and its habits. The best place to release is in the area where it was found, except if the bird risks sustaining the same injury there. Some territorial species, such as kookaburras, must be released back into their own territories or they risk death. See KOOKABURRAS, DID YOU KNOW? page 77. Consider the following before releasing: Is the bird mature enough to fend for itself or does it need further fostering? Does it help itself to food or does it rely on humans to feed it? If the latter, it may need rehabilitation prior to release. Is it territorial? If so, choose the area carefully, preferably releasing it back to its family. Is it nocturnal or diurnal? Diurnal birds should be released in the morning, nocturnal at dusk or night. Does the bird live in a flock or a group? Could it join another group, or could it be released with other birds of the same species? Is the bird migratory? If so, has the group departed for the season? Does the bird live near water? If so, it should be released near a watered area such as a pond, lake, river or the sea. Is the weather suitable for release? If it is pouring rain or blowing a gale, would holding on to the bird for another day be preferable? Is the release spot too noisy, trafficky or visited by too many people? Do not release any birds in a National Park without prior consent from your State Fauna Authority, see page 193. They can also be contacted regarding suggestions for suitable release spots. Wash your hands after handling birds.

## Dead birds

Dead birds can still serve a purpose. Museums can use them for research or display. Birds with bodies intact can be stuffed. Ones which have deteriorated are also useful, as the skeleton can be extracted and mounted. Upon finding a dead specimen, consider taking it to your nearest museum. If the bird is decayed, handle it through a cloth, or lever it with a stick into a bag. Label the bag with the area where the bird was found; on long country trips a rough estimation will do. Also write down the date when found and colour of the eye. If the bird cannot be delivered to the museum immediately, it should be frozen. Wrap it in newspaper, a good insulator. Place in a bag and label, so it does not get defrosted with the steak. It will keep in the deep freeze of the refrigerator for months. When taking it out, wrap it in more newspaper to delay defrosting. Wash your hands after handling birds.

## Euthanasia

Euthanasia is best and most humanely performed at the hands of a veterinary surgeon. Authorities from National Parks and Wildlife NSW advise that it is illegal for the public to destroy native animals. However, there may be extreme circumstances which would create an exception. If the bird is suffering, the condition is irreparable and help is not available, it may be more humane to destroy it on the spot. People found abusing this exception will be penalised. Depending on the size of the bird and the tools available, any of the following are acceptable methods:

Place the bird in a bag into which you have punched several holes. Tie securely to the outlet of the car exhaust pipe and start the engine. Let the car run for a few minutes to allow the fumes to kill the bird. You can deal a sharp blow to the back of the head, with a spade or a heavy rock. Or hold the bird by the legs and strike the head on a hard surface such as a rock.

The bird's neck can be wrung, by holding the head firmly with one hand and the body with the other. Twist the hands firmly in the opposite directions. Do not be alarmed as there may be body movements for some time after death, when using this method. Smaller birds can be placed into a bag and run over with the car.

Before using any of the above procedures, give yourself one last chance to assess the situation. Is euthanasia the only solution; can this bird still be saved at the hands of an expert? Is there no vet anywhere within reach? Could someone else deliver the bird to a vet? Having decided on euthanasia, it is your moral obligation to ensure the job is done with minimal pain to the bird and to the point when the bird is dead. Wash your hands after handling birds.

Important note: These procedures have been written in the humane interest of minimising the suffering of birds.

# KOOKABURRAS AND OTHER KINGFISHERS

## Laughing Kookaburra *Dacelo novaeguineae*

Other names: Kookaburra, Jackass, Giant Kingfisher

Native. Territorial. Active by day.

**Identification**

Average length head to tail 46 cm. Iris: brown. Bill: blackish above, cream below. Legs: brown. Young: similar to adults, with shorter bills and shorter tails; bill: black on both top and bottom.

## Habitat

Lives in family groups. Found mostly in woodlands and timbered districts. Families perch together for the night.
Nest: in a hollow of a tree, or a burrow in a termite mound up in a tree. Has been known to lay eggs in any cavity big enough to accommodate an adult bird.

## Breeding

Breeds mostly from September to January. Nestlings remain in the nest for four to five weeks after hatching. Young are fed by the parents for eight to 13 weeks after leaving the nest.

## Food

Carnivorous. Feeds on insects, invertebrates, small snakes and lizards, rats, mice, eggs and small birds. Seizes food with the bill, killing it by repeatedly pounding on a branch or the ground.

## Emergency food

Adult: canned dog or cat food, mince meat, worms, dog kibble soaked in warm water.
Young: canned dog or cat food. Baby or pet vitamin drops once daily for prolonged feeding. See FEEDING YOUNG BIRDS page 72.

## How to attract

See HOW TO ATTRACT page 63 for general information on nesting places, bird baths and the hazards of cats. A fishpond may attract kookaburras — for the fish, as well as the frogs.

## Handling and transport

Kookaburras are safe to handle provided you stay clear of the bill. See HANDLING page 64 for instructions. Gardening gloves can be worn for extra safety when handling kookaburras. Wash your hands after handling.

Transport in a cardboard box big enough to accommodate the bird comfortably. See TRANSPORT page 65 for more information.

## Did You Know?

*Kookaburras are extremely territorial. The 'laughter' we are so familiar with is their way of advertising their territory, often to be responded to by laughter from another family living nearby. The family unit of these birds has a strict hierarchical structure. Should a bird die, the hierarchy is adjusted and the gap filled. It is therefore extremely important, should you remove a kookaburra from its area, to return it to the same spot within a short time. It will be accepted back by the family only if the hierarchy has not yet been re-adjusted. Young kookaburras can be an exception to this rule. Cases have been noted of a family accepting an unfamiliar baby kookaburra and caring for it as its own.*

# COCKATOOS

Other names: White Cockatoo, Yellow-crested Cockatoo

Native. Active by day.

## Identification

Average length head to tail 49 to 51 cm. Narrow yellow crest on top of head. Crest can be erect or lie flat. Eye: dark brown iris in a white eye-ring. Bill: black. Legs: dark grey. Young: similar to adults; iris: a paler brown.

## Sulphur-crested Cockatoo *Cacatua galerita*

### Habitat
Lives in pairs or flocks. Found in all habitats but prefers timbered areas. Roosts for the night in trees, leaving in the morning to feed. Nest: in a hollow limb or a hole in a tall tree, usually a eucalypt near water. Base of the nest is lined with decayed wood.

### Breeding
Breeds mainly in August to January. Nestlings remain in the nest for five to six weeks after hatching. Young are guarded and fed by the parents for a month or so after leaving the nest.

### Food
A seed-eater. In the morning feeds on seeds on the ground. During the hottest part of the day sits in trees nearby, stripping leaves and bark. Returns to ground-feeding in the afternoon. When the food supply is

exhausted, moves to a new area. Also feeds on seeds of grasses, berries, grain, bulbous roots, insects and their larvae.

**Emergency food**
Adult: mixed parrot seeds, canary seeds, sunflower seeds, fruit, corn, greens.
Young: baby cereal mixed with warm water, mashed ripe fruit, crushed cereal with warm water. Baby or pet vitamin drops once daily for prolonged feeding. See FEEDING YOUNG BIRDS page 72.

**How to attract**
Most people are hesitant to attract these birds as they can be very noisy and may be destructive to both trees and timber on houses. The following natives are a source of food for cockatoos, however they may also attract other birds as well: acacias, *Callitris rhomboidea* (Port Jackson pine, Oyster Bay cypress pine), casuarinas, *Elaeocarpus reticulatus* (blueberry ash), hakeas, *Pittosporum undulatum* (sweet pittosporum, mock orange, native daphne). Choose trees and shrubs best suited to your area. See EXTRA HELP page 193 for sources of more information.
For general information on nesting places, erecting bird baths, and the hazards of cats, see HOW TO ATTRACT page 63.

**Handling**
Cockatoos are safe to handle provided you stay clear of the bill and claws. For general instructions on handling birds, see HANDLING page 64.

**Transport**
Ideally, cockatoos should be transported in cages, as they can chew through most cardboard boxes. A clean, dry metal garbage bin is an alternative. See TRANSPORT page 65 for more information.

**Cockatoos chewing on houses**
It is in this bird's nature to chew idly through the middle of the day. It is also unfortunate to some home owners that western red cedar and some wood-treating oils are so attractive to cockatoos. Apart from the course of their habitual nibbling, they are more likely to chew in an area where a food supply previously offered them has been cut off. It is therefore inadvisable to attract these birds in the first place, by supplementary feeding. Eliminate any ledges or roosting platforms near windows, by blocking them off, cluttering them, making them too slippery or steep to stand on. The birds must be perched in order to be able to chew. If possible, and especially in areas known to have cockatoos, avoid using western red cedar in construction. It is very soft and easily destroyed, and native hardwoods may be just as suitable. Hang strips of coloured paper or tin foil in front of the problem area; their flapping in the breeze may deter the birds from approaching. Using cardboard, plastic or masonite, make a silhouette of a bird of prey such as a hawk; paint it black or roughly to resemble the bird and hang outside the problem area to flap in the breeze, see page 62. Erect a safety screen of chicken mesh or garden netting to protect the troubled area. If all else fails, contact a reliable pest control company to see if they are able to solve your problem. Some major pest control companies sell or hire noise machines to deal with bird problems on a major scale. See BIRDS, How to deter, Page 61.

**Did you know?**

*Some native birds are known for their ability to mimic other birds or other noises. Lyrebirds and bowerbirds are probably the most notable proponents of this skill. Cockatoos and other parrots are also capable of mimicry, but they can only mimic human voices, not the calls of other birds, and only when in captivity.*

# LORIKEETS

## Rainbow Lorikeet *Trichoglossus haematodus*

Other names: Blue Mountain Parrot, Blue Mountain Lory, Bluey, Coconut Lory.

Native. Active by day.

**Identification**

Average length head to tail 29 to 32 cm. Tail can be short or long. Eye: orange-red. Bill: coral. Legs: green-grey.

Young: duller than adults and with a shorter bill; eye: brown; bill: brown or dull reddish-brown.

**Habitat**
Lives in pairs or flocks. Moves around, following the flowering of eucalypts. Roosts for the night in trees, leaving at sunset to feed. Nest: in a hollow limb or a hole in a tree. Usually high up in a eucalypt.

**Breeding**
Breeds mainly in August to January. Nestlings remain in the nest for approximately eight weeks after hatching. Young stay with the parents for some time after hatching, at first returning to the nest to roost.

**Food**
Nectar and pollen eater. Extracts nectar from eucalypt flowers and other natives. Also eats fruit, pollen, berries, blossoms, grain, insects and their larvae. Can be a problem to fruit growers.

**Emergency food**
Adult: baby cereal mixed with warm water, fruit, native flowers, greens, a mixture of water and honey or brown sugar (ratio 9:1). Offer liquid in a container, do not pour down the beak as the bird may choke.
Young: baby cereal mixed with warm water, crushed cereal mixed with warm water. Baby or pet vitamin drops once daily for prolonged feeding. See FEEDING YOUNG BIRDS page 72.

**How to attract**
The following natives are a source of food for lorikeets, however they may also attract other birds: banksias, *Castanospermum australe* (black bean, Moreton Bay chestnut), callistemons, casuarinas, eucalypts, grevilleas, melaleucas, *Schefflera actinophylla* (umbrella tree). Choose those best suited to your area. See EXTRA HELP page 193 for help in choosing. For general information on nesting places, erecting bird baths, and the hazards of cats, see HOW TO ATTRACT page 63.

**Handling and transport**
Lorikeets are safe to handle provided you stay clear of the bill and claws. For general instructions on handling birds, see HANDLING page 64. Gardening gloves can be worn for extra safety when handling lorikeets.

Transport in a cardboard box big enough to accommodate the bird comfortably. See TRANSPORT page 65 for general instructions on transporting birds.

**Did you know?**

*According to some experts, attracting birds by feeding them sugar and water or similar is in the long run damaging. Artificial food is an insufficient diet, and can cause diet-deficiency problems. Birds can become undernourished or obese, bear deformed young or breed out of proportion, thus upsetting the ecological balance. They also accustom quickly to this habit and risk becoming too dependent on the feeder. Needless to say, the effects of having the food supply cut off when the owners are on holidays, or when the house is sold, need no explanation. Persistent birds can also be a problem to the new owners. Use your commonsense in extreme conditions, such as after bushfires or in recently cleared bushland, where supplementary feeding can be helpful.*

# ROSELLAS

## Eastern Rosella *Platycercus eximius*

Other names: Rosella, Rosehill Parakeet
Native. Active by day.

**Identification**
Average length head to tail 30 to 32 cm. Females are duller in colour than males. Eye: brown. Bill: pale grey. Legs: grey.
Young: similar to adults but duller in colour; acquire full colour at about one year.

**Habitat**
Lives in pairs or flocks. Prefers lightly timbered country to dense forests. Frequently seen in parks and gardens.
Nest: in a hollow limb or a hole in a tree, usually a eucalypt. The Eastern Rosella's nest is a few metres off the ground, though it may lay eggs in hollow logs at ground level.

**Breeding**
Breeds mainly in August to January. Nestlings remain in the nest for about 30 days after hatching. Young are in the care of the parents for many months after leaving the nest.

**Food**
Seed-eater. Feeds on seeds both in trees and on the ground. Also eats nectar, nuts, blossoms, berries, insects and their larvae. Can be a problem to fruit growers; however, it helps to eliminate noxious weeds.

**Emergency food**
Adult: budgie seed mix, fruit, greens, crushed cereal, gum leaves.
Young: baby cereal mixed with warm water, mashed ripe fruit.
Baby or pet vitamin daily for prolonged feeding. See FEEDING YOUNG BIRDS page 72.

**How to attract**
The following natives are a source of food for rosellas, however, they may also attract other birds: acacias, banksias, callistemons, eucalypts, grevilleas. Choose trees and shrubs best suited to your area. See EXTRA HELP page 193 for help in choosing. For further information on attracting birds, see HOW TO ATTRACT page 63.

**Handling and transport**
Rosellas are safe to handle, providing you stay clear of their beaks and claws. See page 64 & 65 for general advice on handling and transporting birds.

# CORELLAS AND THE GALAH

## Galah *Cacatua roseicapilla*

Other names: Rose-breasted Cockatoo, Goulie

Native. Active by day.

### Identification

Average length from head to tail 36 cm. Skin colour around eye can vary depending on area where found. Eye: brown in males, pink in females. Bill: horn. Legs: grey. Young: duller in colour than adults, in the first year of life; eye: brown.

### Habitat

Lives in pairs or flocks. Prefers lightly timbered areas. Often frequents parks and suburban gardens. Roosts for the night in eucalypt trees.
Nest: in a hollow limb or a hole in a tree, usually in a eucalypt close to water. After stripping the bark around the nest's entrance, the galah then lines it with leaves.

### Breeding

Breeds mainly in August to November. Nestlings remain in the nest for five to six weeks after hatching. Young stay with the parents for six to eight weeks after leaving the nest.

**Food**
A seed-eater. Feeds mainly on the ground, gathering grass-seeds and other seeds falling from trees and shrubs. Also eats nuts, grain and berries. Can be damaging to cultivated crops; however, it helps in eliminating noxious weeds.

**Emergency food**
Adults: mixed parrot seeds, canary seeds, sunflower seeds, fruit, corn, greens, crushed cereal.
Young: baby cereal mixed with warm water, mashed ripe fruit. Baby or pet vitamin drops once daily for prolonged feeding. See FEEDING YOUNG BIRDS page 72.

**How to attract**
The following natives provide food for Galahs, however they may also attract other birds: acacias, banksias, *Callitris rhomboidea* (Port Jackson pine, Oyster Bay cypress pine), *Elaeocarpus reticulatus* (blueberry ash), *Scaevola calendulacea* (coastal fan-flower). Eucalypts are good roosting trees; choose those best suited to your area. See EXTRA HELP page 193 for help in choosing. See HOW TO ATTRACT page 63 for information on nesting habits, bird baths and cats.

**Handling**
Galahs are safe to handle provided you stay clear of their beaks and claws. See HANDLING page 64 for advice.

**Transport**
Ideally, Galahs should be transported in cages as, like cockatoos, they can chew through most cardboard boxes. A clean, dry metal garbage bin is an alternative, see page 66.

**Did you know?**
*Galahs usually feed in flocks, for safety. In a flock of many feeding birds, there is bound to be at least one whose head is up in time to spot a predator. Their organised life goes beyond their feeding habits. As the adult birds go off to feeding places, the young are left behind in communal nursery groups where they are 'minded' by young Galahs.*

# MAGPIES

## Australian Magpie *Gymnorhina tibicen*

Other names: Black-backed Magpie, Western Magpie, White-backed Magpie, Piping Crowshrike

Native. Territorial. Active by day.

**Identification**
Average length head to tail 44 cm. Depending on area, colour combinations of black and white can vary in pattern. Eye: orange-brown. Bill: grey with a black tip. Legs: black.
Young: similar to adults; dark parts are a mottled grey in the first year, darkening in the second year; bill: dark.

**Habitat**
Lives in flocks or groups of a few, to several hundred, birds. Prefers open woodland for foraging, and dense woodland for roosting. Frequently seen in suburban areas. Nest: bowl-shaped, of twigs and sticks, lined with grass, wool or hair. Placed from six to 16 metres high, in a fork of an outer branch of a tree, usually a eucalypt.

**Breeding**
Breeds mainly in August to October. Nestlings remain in the nest for approximately four weeks after hatching. Young stay with the parents for up to 12 months, being totally dependent on them for the first few months.

**Food**
Omnivorous — will eat both animal and vegetable matter. Feeds on insects, invertebrates, small lizards, frogs, grain, carrion, small birds, and forages around household garbage. Consumes a lot of pest insects.

**Emergency food**
Adult: canned dog or cat food, worms, minced steak.
Young: dog kibble soaked in warm water. See FEEDING YOUNG BIRDS page 72.

**How to attract**
As Magpies can be quite noisy and troublesome during the breeding season, people are reluctant to attract them to their yard. Eucalypts are good nesting trees, though they may attract other birds as well. A bird bath attracts Magpies. See HOW TO ATTRACT page 63 for advice on this, nesting habits and the hazards of cats.

**Handling and transport**
Magpies are safe to handle provided you stay clear of their beaks and claws. See page 64 & 65 for instructions on handling and transporting birds.

**Magpies attacking people**
During the breeding season, Magpies can become very protective towards their young and consider anyone within the nest area an intruder. The problem is a temporary one and the best solution is to avoid the area if possible during breeding time. The bird usually swoops down at people's heads. Wearing a hat is one solution. Children can wear fun hats made from empty ice-cream containers, with large eyes painted on the top using a texta or glued-on cut-outs. More sophisticated solutions include putting up an

umbrella to shelter from the bird. Otherwise wave an umbrella or stick around your head as you walk briskly. In extreme circumstances, the offending bird may be destroyed. Contact your local Police Station for assistance.

**Did you know?**

*The Magpie's territorial structure is well organised. The most successful birds form groups of up to ten birds and choose a small territory which they defend. Their offspring, once independent (at around 12 months), are evicted from the parents' territory. Along with other adult birds that have been evicted from their group, they form a new flock, somewhat nomadic in nature. They roam from place to place in search of food and shelter. In the process, some birds leave the flock to join other small groups of settled birds, or form a new group, eventually dispersing the original flock.*

# CURRAWONGS

## Pied Currawong *Strepera graculina*

Other names: Currawong, Black Magpie, Chillawong
Native. Active by day.

## Identification

Average length head to tail 45 to 46 cm. Tip of flight feathers are white, concealed in wings and more evident in flight. Eye: yellow. Bill: black. Legs: black.
Young: similar to adults but duller; eye: darker than those of adults.

## Habitat

Lives in pairs or flocks. Prefers forest areas, flocking in tall timber where it roosts for the night. Frequents suburban gardens, especially in winter.
Nest: bowl-shaped, of twigs and sticks, lined with grass or rootlets. Placed from seven to 20 metres up in an outer branch of a fork of a tree, usually a eucalypt.

## Breeding

Breeds mostly in September to January. Nestlings remain in the nest for approximately four weeks. Young are fed by adult birds for approximately two months after leaving the nest.

## Food

Omnivorous — will eat both animal and vegetable matter. Feeds on insects, young birds, fruit, berries, and carrion. Can be damaging to fruit growers.

## Emergency food

Adult: canned dog or cat food, worms, fruit, minced steak.
Young: dog kibble soaked in warm water. See FEEDING YOUNG BIRDS page 72.

## How to attract

The following natives provide food for currawongs. However, they may attract other birds as well: acacias, *Elaeocarpus reticulatus* (blueberry ash), *Pittosporum undulatum* (sweet pittosporum, mock orange, native daphne), *Santalum acuminatum* (sweet quandong). Eucalypts and *Pittosporum undulatum* are good nesting trees for currawongs, as well as other birds. Choose those best suited to your area. See EXTRA HELP page 193 for help in choosing. See HOW TO ATTRACT page 63 for erecting a bird bath and a caution about cats.

## Handling and transport

Currawongs are safe to handle provided you stay clear of their beaks and claws. See page 64 & 65 for general advice on handling birds, and transporting them.

# WATTLEBIRDS

## Red Wattlebird *Anthochaera carunculata*

Other names: Gillbird

Native. Territorial. Active by day.

### Identification
Average length head to tail 32 to 35 cm. The largest mainland honeyeater. Can be identified by the pinkish-red wattle behind the eye. The wattle increases with age, reaching up to two centimetres in mature birds. Eye: chestnut-red. Bill: black. Legs: brown-grey. Young: similar to adults, with a small, or no, wattle.

### Habitat
Lives in small groups, forming flocks when searching for food. Frequents forest areas, though single birds are often seen in surburban gardens, feeding on nectar-producing shrubs. Flocks move in search of food.
Nest: cup-shaped, roughly constructed from grass and twigs; lined with soft material, feathers or wool, and placed in a bush or tree two to ten metres above ground.

### Breeding
Breeds mainly in July to December. Nestlings remain in the nest for approximately two weeks. Young are cared for by the male bird for some time after leaving the nest.

### Food
A honeyeater. Feeds on nectar, fruit and insects including spiders. Particularly fond of banksias.

### Emergency food
Adult: soft ripe fruit, native flowers, a mixture of water and honey or brown sugar (ratio 9:1 for larger birds, 5:1 for smaller ones). Offer liquid in a container. Dip the bird's beak in and wait for it to drink. Never pour liquid down the throat as the bird may choke.
Young: soft fruit, dog kibble soaked in warm water, baby cereal mixed with warm water. See FEEDING YOUNG BIRDS page 72.

### How to attract
The following natives provide food for wattlebirds; however, they may also attract other birds: *Angophora cordifolia* (dwarf apple myrtle, heart leaf myrtle), banksias, callistemons, *Calothamnus, Castanospermum australe* (black bean, Moreton Bay chestnut), eucalypts, grevilleas, hakeas, *Anigosanthos* (kangaroo paws), melaleucas, *Pittosporum undulatum* (sweet pittosporum, mock orange, native daphne), *Schefflera actinophylla* (umbrella tree), *Stenocarpus sinuatus* (firewheel tree), telopeas. *Castanospermum australe* is a good tree for shelter. Choose those best suited to your area. See EXTRA HELP page 193 for help in choosing. See page 63 for advice on bird baths and cats.

**Handling and transport**
Wattlebirds are safe to handle. Their sharp claws are the main concern and should be covered or kept away from you. See HANDLING page 64 for advice on handling through a cloth or with bare hands. See page 65 for instructions on transporting birds.

**Did you know?**
*The Red Wattlebird is now protected by law. However, not too long ago these birds were hunted as game birds.*

# HONEYEATERS

## New Holland Honeyeater *Phylidonyris novaehollandiae*

Other names: White-eyed Honeyeater, Yellow-winged Honeyeater, White-bearded Honeyeater

Native. Active by day.

## Identification

Average length head to tail 16 to 18 cm. White stripes on the throat area give the bird a bearded look. Iris: white. Bill: black. Legs: black. Young: similar to adults but duller; mainly dull brown body marked with yellow and grey. Iris: brownish or greyish.

## Habitat

Lives in pairs or flocks. Prefers open woodland and dense scrub and coastal districts. Frequents gardens in search of food.
Nest: cup-shaped, made of strips of bark, grasses and twigs; lined with soft plant material, usually from banksia flowers. Placed in a shrub or small tree, rarely more than three metres above ground.

## Breeding

Breeds mainly in spring, though at any time if conditions are favourable. Nestlings remain in the nest for approximately 16 days after hatching. Young are fed for quite some time by the parents, after leaving the nest.

## Food

A honey eater. Feeds on nectar, fruit and insects. Particularly fond of banksias and grevilleas.

## Emergency food

Adult: soft ripe fruit, native flowers, a mixture of water and honey or brown sugar (ratio 4:1). Offer liquid in a container. Do not pour down the throat as the bird may choke. Young: soft fruit, dog kibble soaked in warm water, baby cereal mixed with warm water. See FEEDING YOUNG BIRDS page 72.

## How to attract

The following natives are a source of food for honeyeaters, however, they may also attract other birds: *Angophora cordifolia* (dwarf apple myrtle, heart leaf myrtle), banksias, callistemons, *calathamnus* , correas, eremophilas, eucalypts, grevilleas, *Anigosanthos* (kangaroo paws), hakeas, *Lambertia formosa* (honey flower, mountain devil), melaleucas, *Schefflera actinophylla* (umbrella tree), telopeas. Choose those best suited to your area. See EXTRA HELP page 193 for help in choosing. See page 63 for advice on bird baths and cats.

## Handling and transport

This small bird is safe to handle. See page 64 for general instructions on handling birds. See page 65 for advice on transporting birds.

# MINERS

## Noisy Miner *Manorina melanocephala*

Other names: Snakebird, Micky, Noisy Mynah, Soldier-bird, Squeaker

Native. Territorial. Active by day.

## Identification

Average length head to tail 25 to 28 cm. Has a yellow patch of bare skin behind the eye area, merging with the eye. Eye: brown. Bill: yellow. Legs: ivory. Can be distinguished from the Indian Myna by the grey body. Indian Mynas are more brown.
Young: similar to adults, but more brown.

## Habitat

Lives in groups of a few to several birds. Prefers open timber country. A frequent visitor of suburban gardens.
Nest: cup-shaped, made of twigs and grasses bound with cobwebs. Lined with soft plant material, wool or hair. Placed in an outer branch of a tree, three to 20 metres above ground.

## Breeding

Breeds mainly in June to January. Nestlings remain in the nest for approximately 15 days. Young return to the nest at night for a few days after leaving the nest. They remain with the family group, though may eventually move to a new group.

## Food

A honey eater. Feeds on nectar, berries, small insects and small invertebrates. Can be damaging to fruit growers.

## Emergency food

Adults: ripe fruit, native flowers, a mixture of water and honey or brown sugar (ratio 5:1), baby cereal mixed with warm water. Offer liquid in a container. Dip the bird's beak into the liquid and wait for it to drink. Never pour liquid down the beak as the bird may choke.
Young: soft fruit, dog kibble soaked in warm water. See FEEDING YOUNG BIRDS page 72.

## How to attract

The following natives provide food for Noisy Miners, however, they may attract other birds as well: banksias, *Calothamnus* , callistemons, eucalypts, grevilleas, hakeas, melaleucas, *Schefflera actinophylla* (umbrella tree), telopeas. *Castanospermum australe* (black bean, Moreton Bay chestnut) is a good shelter tree, especially from the midday heat. Choose ones suited to your area. See EXTRA HELP page 193 for help in choosing. See HOW TO ATTRACT page 63 for advice on bird baths and cats.

## Handling and transport

Miners are safe to handle. Their sharp claws are the main concern and should be covered or kept away from you. See page 64 for instructions on handling birds through a cloth or with bare hands. Transport: page 65.

### Did you know?

*Noisy miners live in tight family groups and tackle anything in a communal fighting spirit. A group of them can often be seen hassling much larger birds with their constant chirping, squawking, and enough nagging to send any intruder packing. Meanwhile, back at the nests, the male birds are so attentive to the young, that the newborn have no idea who their real fathers are.*

# MYNA

## Indian Myna *Acridotheres tristis*

Other names: Common Myna

Introduced. Active by day.

**Identification**
Average length head to tail 23 to 24 cm. Has a large patch of yellow bare skin behind and around the eye. A white patch on the wing becomes more visible in flight. Eye: brown. Bill: yellow. Legs: yellow. Can be distinguished from Noisy Miners by the brown body. Noisy Miners are grey.
Young: similar to adults but duller.

**Habitat**
Lives in communal roosts. Being a scavenger, lives anywhere where there is an abundance of food. Found mainly in cities, in parks and suburban gardens. Roosts in large numbers in dense trees, with a liking for palms.
Nest: competes with native birds, successfully taking over existing nests in tree hollows. Otherwise, makes an untidy nest of grasses, placed in tree hollows or in crevices such as in buildings, or in thick vegetation.

**Breeding**
Breeds mainly in October to March. Nestlings remain in the nest for 20 to 24 days after hatching.

**Food**
Originally insectivorous, in Australia it has become omnivorous and a scavenger. Will adjust to the food available in the area, and will eat virtually anything. Prefers to feed on the ground. Frequents backyards, parks, and rubbish dumps in search of food.

**Emergency food**
Adult: canned dog or cat food, minced meat.
Young: dog kibble soaked in warm water. See FEEDING YOUNG BIRDS page 72.

**How to attract**
It is not a bird one should try to attract to the garden, as it can make quite an undesirable guest. It is noisy and may build a nest in your roof. It also competes with native birds for food and shelter. Without going to any trouble to attract it, it is found in most gardens.

**Handling and transport**
Mynas are safe to handle. Their claws are moderately sharp and should be covered or kept away from you. See page 64 for instructions on handling birds through a cloth or with bare hands. See page 65 for instructions on transporting birds.

**Did you know?**
*Though seen frequently in our parks, yards and streets, this bird is not native to Australia. Being insectivorous, it was introduced from South-East Asia in the 1860s to help keep down some insect populations, which were in plague proportions.*

# SILVEREYES

## Eastern Silvereye *Zosterops lateralis*

Other names: White-eye, Eastern Silvereye, Silvey, Grey-breasted Silvereye

Native. Migratory. Active by day.

### Identification

Average length head to tail 10 to 12 cm. Colour combination of greens, greys and yellows can vary regionally. Eye: brown, circled by a ring of white feathers. Bill: brown.

Legs: brown.
Young: similar to adults, with fluffier feathers.

### Habitat

Lives in pairs or flocks. Found in forest country as well as surburban gardens. Moves in groups in seach of food.
Nest: cup-shaped, made of fine grasses or hair, and bound with cobweb. Lined with soft plant material or hair. Placed on a horizontal outer branch in a bush, from one to four metres above ground.

### Breeding

Breeds mainly in August to January. Nestlings remain in the nest for up to two weeks after hatching.

### Food

Insectivorous, but with a liking for fruit. Feeds on insects found on tree leaves, nectar, seeds, berries, and soft fruit. Can be a problem to fruit growers.

### Emergency food

Adults: soft fruit, a mixture of water and honey or brown sugar (ration 4:1). Offer water in a dish. Do not pour down the bird's beak as the bird may choke.
Young: soft ripe fruit, dog kibble soaked in warm water. See FEEDING YOUNG BIRDS page 72.

### How to attract

The following natives attract insects during the flowering period; the insects in turn may attract insect-eating birds such as the Silvereyes: *Angophora cordifolia* (dwarf apple myrtle, heart leaf myrtle), acacias, casuarinas, eucalypts, grevilleas, hakeas, melaleucas, *Schefflera actinophylla* (umbrella tree), *Solanum aviculare* (kangaroo apple), *Pittosporum undulatum* (sweet pittosporum, mock orange, native daphne), and any native fruit trees such as native plums. *Callitris rhomboidea* (Port Jackson pine, Oyster Bay cypress pine) is a good shelter and nesting tree. Choose those best suited to your area. See EXTRA HELP page 193 for help in choosing. See page 63 for advice on bird baths and cats.

### Handling and transport

This small bird is safe to handle. See page 64 for instructions on handling birds through a cloth or with bare hands.

Transport in a small cardboard box or an empty ice-cream container. See page 65 for general instructions.

### Did you know?

*A migrating bird is one which changes area at the same time every year, forming a clear pattern. It migrates to avoid severe weather conditions or for breeding purposes. A migrating bird will breed in one area and spend most of its non-breeding time in another area, returning to the same breeding area at the same time every year. Bird migration is still under close investigation, as many questions, such as how they find their way, remain unanswered.*

# PARDALOTES

## Spotted Pardalote *Pardalotus punctatus*

Other names: Pardalote, Diamondbird, Spotted Diamondbird

Native. Active by day.

### Identification

Average length head to tail 9 cm. Can be identified by pronounced white spots on the forehead, wings and tail. Females are duller in colour than males, with the spots more creamy than white. They also lack the yellow colour on the throat. Eye: grey. Bill: black. Legs: flesh-brown.
Young: similar to adult females, but duller.

**Habitat**
Lives in pairs or flocks. Found in forests and surburban gardens. Flocks flit from tree to tree in search of insects.
Nest: excavates on the side of a cliff or in the ground. Forms a tunnel, ending with a nest chamber, lined with grass or soft plant material. Rarely nests in a hole in a tree.

**Breeding**
Breeds mainly in August to January. Nestlings remain in the nest for 21 to 25 days.

**Food**
Insectivorous. Forages in trees in flocks, searching for small insects and small invertebrates, such as caterpillars, moths, spiders, thrips and beetles.

**Emergency food**
Adult: dog kibble soaked in warm water, sieved hard-boiled eggs, insects.
Young: dog kibble soaked in warm water. See FEEDING YOUNG BIRDS page 72.

**How to attract**
The following natives attract insects during the flowering period, which in turn may attract insectivorous birds such as Pardalotes: acacias, *Angophora cordifolia* (dwarf apple myrtle, heart leaf myrtle), eucalypts, grevilleas, hakeas, maelaleucas. *Pittosporum undulatum* (sweet pittosporum, mock orange, native daphne) is a good shelter tree. Choose ones best suited to your area. See EXTRA HELP page 193 for help in choosing. Erect a bird bath, preferably near vegetation. See page 63 for instructions on erecting a bird bath, and a warning about cats. If you have any Pardalotes nesting on your property, cat-proof the area of the nest entrance. Cover with a mesh wide enough for the birds to enter, but small enough to stop the cat.

**Handling and transport**
This small bird is safe to handle. See page 64 for instructions on handling birds through a cloth or with bare hands. See page 65 for instructions on transporting birds. A small cardboard box or an empty ice-cream container are suitable for Pardalotes.

**Did you know?**

*Why do birds breed in spring? Their natural instinct guides them to bring young into the world at this time, when there is an abundance of food. However, their built-in instincts are more sophisticated than that. The birds can also control their breeding to conform with extreme weather conditions. In certain parts of Australia which are prone to droughts, birds choose to curb their breeding or stop it should extreme weather conditions continue for extended periods. To compensate for the loss, during favourable conditions they extend their breeding season.*

# CUCKOOS

## Pallid Cuckoo *Cuculus pallidus*

Michael Morcombe

Other names: Brainfever-bird, Grasshopper Hawk, Mosquito Hawk, Scalebird

Native. Migratory. Active by day.

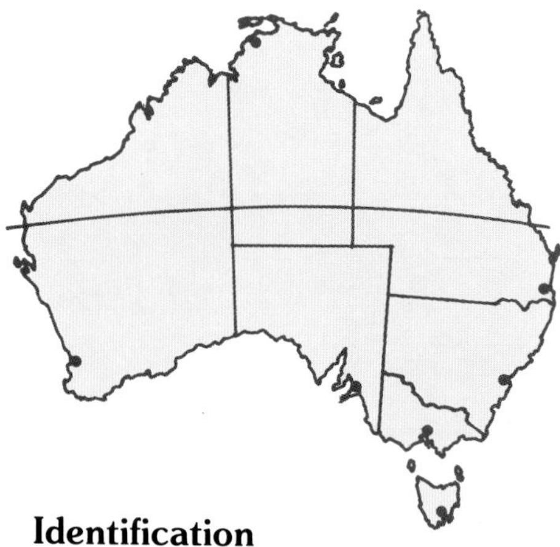

## Identification

Average length head to tail 30 to 33 cm. Body is all grey when the plumage is new, darkening more to a brown with older plumage. Eye: brown with a yellow eye-ring. Bill: dark olive. Legs: olive-brown. Young: heavily mottled shades of cream and brown in young birds, gradually decreasing to a more uniform colour as the bird matures.

## Habitat

Usually seen as a single bird. Found throughout Australia, though shows a preference for coastal areas over dry ones. Migrates north in winter, south in spring.
Nest: does not construct a nest. Lays eggs in nests of other birds. Host birds are chosen usually because of the similarity of their eggs to those of the cuckoo.

## Breeding

Breeds mainly in September to January. Having deposited the eggs, the Pallid Cuckoo leaves the incubating and rearing to the host bird. Rate of growth of the young is usually faster than that of the host's young.

## Food

Insectivorous. Prefers to feed on the ground, looking for invertebrates, caterpillars being the favourite. Other foods include grasshoppers and beetles.

## Emergency food

Adult: worms, soft fruit, caterpillars, moths, any insects. Young: dog kibble soaked in warm water. See FEEDING YOUNG BIRDS page 72.

## How to attract

Eucalypts and acacias make good roosting trees; however, they may attract other birds as well. See page 63 & 64 for information on bird baths and cats.

## Handling and transport

Cuckoos are safe to handle. Their claws are not dangerously sharp and their bites not dangerous. See page 64 for instructions on handling with a cloth or with bare hands. See page 65 for instructions on transporting birds.

## Did you know?

*The reproduction method of this bird is parasitic. It does not build a nest, but deposits its eggs in the nest of another bird. The cuckoo usually chooses a host whose eggs are like those of the cuckoo. There are approximately 80 species of suitable hosts, the honeyeaters being the most frequent victims. Having laid the eggs, the cuckoo departs, leaving the host bird not only to hatch the eggs, but to foster the young as well. In some cases an odd combination arises, with a tiny foster-mum feeding a 'baby' twice her size.*

# OWLS

## **Boobook Owl** *Ninox novaeseelandiae*

Michael Morcombe

Other names: Spotted Owl, Mopoke, Southern Boobook

Native. Nocturnal.

### Identification
Average length from head to tail 30 to 35 cm. Colour combinations of brown and white vary regionally. Eye: yellow. Bill: blue-grey. Feet: blue-grey.
Young: similar to adults, with a brown eye.

### Habitat
Lives singly or in pairs. Found in all habitats, showing no preference for any in particular, as long as perches are available. Roosts by day in dense foliage, in caves if no trees are available, or anywhere it can perch.
Nest: in a hollow limb or a hole in a tree, lined with wood-dust or bark. Placed from one to 20 metres above ground.

### Breeding
Breeds mainly in September to January. Nestlings remain in the nest for 37 to 44 days after hatching. Young stay with the parents for several months after leaving the nest.

### Food
Carnivorous and insectivorous. Feeds on small birds and small mammals such as mice. Also preys at night for flying insects, hence its frequent presence near house or street lights at night.

### Emergency Food
Adults: any of the following may be offered after dark: canned dog or cat food, chopped meat with a pinch of grainy sand or dirt, and calcium powder if available. Will readily eat live mice.
Young: chopped meat with a pinch of grainy sand or dirt, with calcium powder if available. See FEEDING YOUNG BIRDS page 72.

### How to attract
Tree hollows are potential nesting places for owls and many other birds. Where possible, do not 'tidy' the trees by removing any dead branches. *Syzygium leuhmannii* (small-leaf lillypilly) is a good roosting tree. If the outside light is left on, it will attract insects which in turn may attract owls as well as other nocturnal insectivorous birds. (This does not apply to lights with yellow globes). See page 63 & 64 for a warning on cats.

### Handling and transport
Owls are safe to handle provided you stay clear of their claws which are extremely sharp. See page 64 for instructions on handling through a cloth or with bare hands. See page 65 for instructions on transporting birds.

# FROGMOUTHS AND NIGHTJARS

## Tawny Frogmouth *Podargus strigoides*

Other names: Mopoke

Native. Nocturnal.

### Identification

Average length head to tail 40 to 46 cm. Females are slightly smaller than the males. Resembles a broken-off branch, making it difficult to detect in a tree. Eye: orange-yellow. Bill: brown. Legs: dull brown.

Young: similar to adults, with fluffy plumage.

### Habitat
Lives singly, in pairs, or in family groups. Found wherever there are trees, including gardens and city parks. Roosts by day on a tree branch, itself resembling a branch. Nest: a loose platform of twigs with a slight depression in the middle. Lined with green leaves and placed horizontally in a forked branch of a tree, usually a eucalypt. From five to 15 metres above ground.

### Breeding
Breeds mainly in August to December. Nestlings remain in the nest for 25 to 35 days after hatching. Young depend on the parents for food for one to two weeks after leaving the nest, though they remain with the parents for several months after this.

### Food
Insectivorous and carnivorous. Forages on the ground looking for insects and spiders. Often found near street lights, preying on flying insects. Capable of swooping on many insects in a swarm. Also feeds on mice.

### Emergency food
Adult: any of the following may be offered after dark: canned dog or cat food, chopped meat with a pinch of grainy sand or dirt and calcium powder if available. Will also accept live mice. Hint: pat lightly on the back of the head, and it should gape its mouth open. Young: canned dog or cat food, chopped meat with a pinch of grainy sand or dirt and calcium powder if available. See FEEDING YOUNG BIRDS page 72.

### How to attract
Eucalypts, dead or alive, are good roosting trees for Frogmouths as well as many other birds. If the outside light is left on at night, it will attract insects which in turn may attract Frogmouths as well as other nocturnal insectivorous birds. (This does not apply to lights with yellow globes). See page 63 & 64 for warning on cats.

### Handling and transport
Frogmouths are safe to handle provided you stay clear of their bills. Their claws are moderately sharp. See page 64 for general instructions on handling birds. See page 65 for instructions on transporting birds.

### Did you know?
*Though resembling an owl, the Tawny Frogmouth is not one. It can be easily differentiated from owls by its wide beak and weak feet. An owl, being a bird or prey, has strong claws. The Tawny Frogmouth is extremely efficient when it comes to camouflage. When sitting on a branch, its feathers are hard to distinguish from the tree bark. It assumes a stick-like posture, closes its eyes and points the bill upwards, making it impossible to detect.*

# SWALLOWS

## Welcome Swallow *Hirundo neoxena*

Other names: Swallow, House Swallow, Australian Swallow

Native. Active by day.

### Identification
Average length head to tail 15 cm. Can be identified by the long, deeply forked tail. Eye: dark brown. Bill: black. Legs: dark brown. Young: similar to adults, but duller in colour.

### Habitat
Lives in flocks. Likes both countryside and built-up areas. Though a migratory species, in some areas chooses to remain all year round. Often seen near houses, car parks, bridges and mineshafts.
Nest: cup-shaped, made of mud pellets and vegetation, lined with grass and soft material. Bound to a hard vertical surface such as a wall, preferably one with an awning.

### Breeding
Breeds mainly in August to December. Nestlings remain in the nest for up to 25 days after hatching. Young are dependent on the parents for approximately three weeks after leaving the nest.

**Food**
Insectivorous. Feeds in mid-air, catching insects such as flies, midges and moths.

**Emergency food**
Adults: canned dog or cat food, soft fruit, sieved hard-boiled egg. Young: any insects attracted to the light such as moths; canned dog or cat food. See FEEDING YOUNG BIRDS page 72.

**How to attract**
As these birds feed mainly on insects flying in mid-air, very little can be done in terms of attracting them by providing food. Any ledges

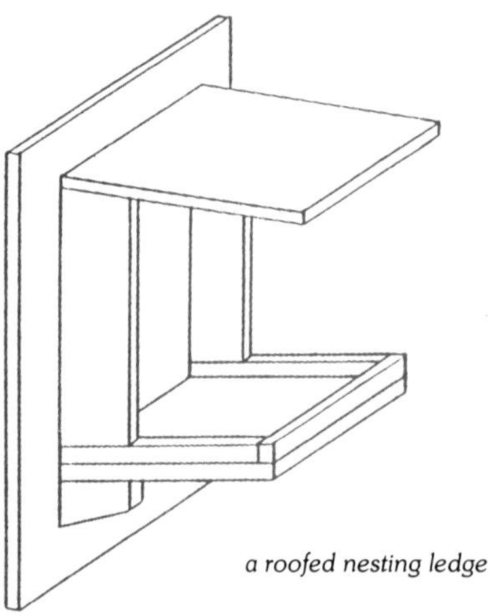
*a roofed nesting ledge*

under cover may attract them to nest; however, other birds such as pigeons find such places equally attractive. If there are swallows in your area and you would like a family of them in your yard, you can construct a nesting ledge for them. There is no guarantee that swallows will take up residence, or that other birds won't. See page 63 & 64 for information about bird baths and cats.

**Handling and transport**
This small bird is safe to handle. See page 64 for general instructions on handling.

See page 65 for instructions on transporting birds.

**Relocating a swallow's nest**
Should a swallow's nest fall, or be erected in a place where it is troublesome or in danger, it can be relocated. First, choose a new site for the nest on any sheltered vertical surface, away from cats' reach. Unless the birds are in danger, leave the relocating procedure till the young have left the nest. When moving with the young, gently remove the babies or eggs from the nest. Place in a safe, warm place such as an empty ice-cream container lined with cloth. Handle as little as possible.

If the nest is not down, loosen it by wedging something sharp such as a knife or spatula between the nest and the wall. The nest should be removed in one piece. Wet the edges of the nest lightly with water to soften, then press firmly into position on the new surface. Hold for 10 to 15 minutes. If the nest resists sticking, try sticking the edges with Blu-tack.

Another method of relocation is using a pouch. Cut a triangular shape out of material or garden netting, slightly larger than the nest. Fasten the triangle to the new nest site, widest side up and the apex pointing down, so that a pouch is formed between the triangle and the vertical surface. Wet the edges of the nest slightly with water, and slip into the pouch. Press the edges of the nest onto the wall surface and hold for a few minutes. Gently place the eggs or the young into the nest and leave. Observe from a distance to ensure that the parents have returned to attend to the eggs/young.

If the nest is relocated close to

the original site and the procedure carried out swiftly, the parents should be back to attend to the young on the new site. However, if after considerable observation you believe the young abandoned, see FEEDING YOUNG BIRDS page 72 and SWALLOWS, EMERGENCY FOOD page 110. Ring for help, see EXTRA HELP page 193, re fostering the birds. Wash your hands after handling the nest or the birds.

**Did you know?**

*All birds groom or 'preen' themselves. Preening helps to realign the feathers, keeping them waterproof and in top condition, which is necessary for efficient flight. Preening is done daily and preferably after a bath. Using the beak, each feather is nibbled at, removing any dirt; it is then realigned and coated with an oil for waterproofing. The oil is secreted from a gland situated at the base of the tail. Most birds preen themselves, though some indulge in mutual preening.*

# GULLS

## Silver Gull *Larus novaehollandiae*

Other names: Seagull, Red-billed Gull, Red-legged Gull
Native. Active by day.

## Identification
Average length head to tail 40 to 42 cm. Eye: white with a red eye-ring. Bill: bright red. Legs: bright red.
Young: body is mottled with brown and grey; eye: brown; bill: brown; legs: brown. As the bird matures, eye becomes white, bill and legs become more red.

## Habitat
Lives in flocks. Abundant along the coast and along the beaches and in rubbish dumps in cities. Flies daily to roosting places on isolated islands and areas with very shallow water.
Nest: a shallow depression in the ground, lined with grass or dry seaweed. Also nests in low shrubs or grasses.

## Breeding
Breeds at different times of the year, depending on area. Nestlings remain in the nest area for three to four weeks after hatching. Young stay within the parents' territory for a further two weeks. If they stray, they are liable to be injured or killed by other gulls.

## Food
Omnivorous, and a scavenger. When at sea, feeds on small fish, plankton, and scavenges behind fishing boats. Near the beach feeds on washed-up seaweed, crustaceans and handouts from the public. In cities feeds on rubbish in dumps, and scavenges off the public. Also eats eggs of other seabirds.

## Emergency food
Adult: canned dog or cat food, minced meat with a pinch of calcium powder if available, chopped fish.
Young: canned dog or cat food, minced meat with a pinch of calcium powder if available, chopped fish. See FEEDING YOUNG BIRDS page 72.

## How to attract
You cannot attract gulls into your yard unless you live in an area where they flock, such as near the beach or a park. But, as they live in flocks, it is not advisable to attract them or you will end up with a yard full of gulls. One consolation is that they probably would not stay long, as being an opportunistic feeder, something better always comes along sooner or later to attract them to another area. When in 'their' territory, attracting them is very simple. Any biscuit, chip or a piece of bread thrown up in the air, will soon catch their attention.

## Handling and transport
Gulls are safe to handle provided you stay clear of their beaks and claws. See page 64 for instructions on handling birds through a cloth or with bare hands.
See page 65 for instructions on transporting birds.

**Did you know?**

*That one Silver Gull can cause major damage to a jumbo jet? By flying into its engine, it has the potential to cause its malfunction. Deterring birds from airport areas is a major safety procedure. The grass has to be cut low to stop the birds from foraging for food. All garbage has to be removed and stored off the premises. Only trees unsuitable for roosting, such as swamp-pines, are planted within close proximity to the runway. There is a noise machine on hand, to deter any potential stayers. Though gulls are the main concern, all birds, even sparrows, are a potential hazard at an airport.*

# DUCKS

## Grey Teal *Anas gibberifrons*

Other names: Slender Teal, Wood Teal
Native. Active by day.

### Identification

Average length head to tail 40 to 47 cm. Females are slightly smaller than males. Iris: brown to red, depending on age and sex. Bill: black. Legs: black.
Young: similar to adults in colour, with a slight variance in the pattern; iris: brown.

## Habitat
Lives in pairs or flocks. Found throughout Australia in any watery habitat, including both salt and fresh water.
Nest: made of grass and lined with down. Placed anywhere: on the ground, in crevices, rabbit burrows, in tree holes or in reeds.

## Breeding
Breeds anytime of the year, provided water level is high enough, which is usually influenced by rain. Nestlings leave the nest almost immediately after hatching. Young stay with the parents for four to six weeks after leaving the nest.

## Food
Basically a vegetarian, though consumes a variety of food. Nibbles on plants growing along water's edge, as well as in the water. In the water, feeds on shrimps, mussels and small aquatic animals. On land, feeds mainly on insects such as dragonflies, beetles and their larvae.

## Emergency food
Adult: the following should be served dry: fowl pellets, budgie seed mix, crushed dog biscuits or kibble. The following should be served in water: greens, millet seeds. Water must be available at all times.
Young: greens floating in shallow water. Water must be available at all times.

## How to attract
The only way to end up with ducks in your yard is if you live near water, such as a lake or a lagoon. Quite unintentionally, people in urban areas, who have pools, have ended up with ducks in their yards. However, they are only transitory residents. If you find the company of ducks relaxing, enquire at your local council about creating an artificial duck pond at your local park. Keep in mind that not all residential areas lend themselves to such a project.

## Handling and transport
Ducks are safe to handle as their claws are only moderately sharp. See page 64 for instructions on handling birds through a cloth. When handling with bare hands,

grab the duck with both hands, one on each side of its body, thus immobilising the wings and the feet. Another method is to place the duck under your arm, holding onto the top of its legs with one hand and the back of the neck with the other. Gloves can be worn for extra safety.
See page 65 for instructions on transporting birds.

**Ducks in the pool**
A family of ducks may accidentally end up in your pool. Do not panic, as they cannot harm you or the pool. The first thing to attend to are your pets. Lock them up if possible so they do not frighten and scatter the ducks. The next is the pool filter. Tiny ducklings may be sucked up into the skimmer box. If possible, do not run the pool filter; otherwise place a net over the skimmer box. If the presence of these ducks does not inconvenience you, the best solution is to leave them alone. They should shift in a matter of days, as your pool could not sustain them for a long time. The other alternative is to move them. Their capture is not easy and may result in the parents flying away in the process, leaving you with the ducklings, so care must be taken.

You will need a place within your yard with a door, such as a tool shed, where you can trap the parents. Gently scoop the ducklings out of the pool with the pool scoop, and place them in a box. When you have captured the lot, place the box with the ducklings in the area where you can trap the parents. Open the door, and let the parents listen to the cries of the babies. They should respond by walking into the area with the ducklings. At that point the door should be shut to trap them inside.

Once in a smaller area, you should attempt to catch the parents. Throw a towel or a cloth over the ducks and scoop them with the cloth into a box. Should the parents fly off at any point, use the box with the ducklings as 'bait' to lure them back to the area. Leave the ducklings in the yard in the box, so they can be heard by the parents. Having captured the entire family, release them in a quiet, watered area such as a pond, lake or river. Ring your State Fauna Authority for possible release spot suggestions, see page 193.

Should you be left with a misplaced duckling after the release, try to unite it with the family. If that is not possible, or if all the ducklings have been abandoned by the parents, the duck/s will need to be fostered. See DUCKS, EMERGENCY FOOD page 114 to feed the ducks. Ring for help, see EXTRA HELP page 193 re fostering the duck/s. Wash your hands after handling ducks.

**Did you know?**
*What do birds do for drinking water when they fly out to sea for months at a time? They drink salt water. Their bodies are equipped with special glands which take the salt content out of sea water.*

# STARLINGS

## Common Starlings *Sturnus vulgaris*

Other names: Starling, European Starling, English Starling
Introduced. Active by day.

**Identification**
Average length head to tail 20 to 25 cm. New plumage is spotted, spots eventually wearing off. Eye: dark brown. Bill: yellow when breeding, dark brown when non-breeding. Legs: reddish-brown. Young: body is dull brown all over. Spots appear on new plumage as the bird approaches adulthood.

## Habitat

Lives in groups or flocks. Found both in the bush as well as in the cities. Roosts for the night in trees, in noisy communal flocks.
Nest: competes with native birds, successfully taking over existing nests in tree hollows. Otherwise, makes an untidy nest of twigs and rubbish, lined with softer materials. Placed in a tree hole, a rock crevice or between the roof and ceiling of houses.

## Breeding

Breeds mainly in August to January. Nestlings remain in the nest for approximately three weeks after hatching. Young are fed by the parents for a short time after leaving the nest.

## Food

Omnivorous. Prefers to feed in groups on the ground, looking for insects and seeds. Also feeds on insects and fruit in trees and shrubs. Can be a pest to fruit growers, though consumes some pest insects in return.

## Emergency food

Adult: canned dog or cat food, minced meat.
Young: dog kibble soaked in warm water. See FEEDING YOUNG BIRDS page 72.

## How to attract

A word of caution. Do not encourage this bird into your yard, as it may nest in your roof which is a troublesome situation. It also roosts in noisy communities and competes with native birds for food and shelter.

## Handling and transport

This bird is safe to handle. See page 64.

See page 65 for advice on transporting birds.

## Starlings nesting in the roof

If you have starlings nesting in your roof, it is advisable to have them removed. Apart from being noisy, when the young leave the nest other problems are created. The nest can become a fire hazard. It also contains red blood-sucking mites which are left behind after the host birds have gone. The mites descend into your living area and embed themselves into the linen. They can give you an itch and are potential carriers of disease. You may choose to call out a reliable pest exterminator to deal with the problem, or tackle it yourself. Depending on your conscience and the severity of the problem, it may be more humane to destroy the nest once the young have left. Once empty, remove the nest from the roof area and burn it. Spray the roof area around where the nest was located with a domestic insect surface spray. Examine the area to locate where the starlings may have entered the roof; this is important to ensure the problem is not repeated. Block off and seal any holes or possible entry points to the roof. Wash your hands after handling the nest.

# PIGEONS

## Domestic Pigeon *Columba livia*

Other names: Rock Pigeon, Feral Pigeon, Rock Dove

Introduced. Feral. Active by day.

### Identification
Average length head to tail 33 to 34 cm. Interbreeding with other domestic show pigeons has resulted in some odd colour combinations. Eye: orange. Bill: black. Legs: reddish-brown. Young: similar to adults but with a smaller wattle above the bill; eye: brown.

### Habitat
Lives in colonies. Found mainly in urban environment, in every town and city in Australia. Roosts on stable surfaces such as ledges of buildings and other man-made structures.

Nest: made of sticks, placed on any flat surface. In urban environment placed on ledges of buildings, for shelter; otherwise on side of cliffs or in tree holes.

### Breeding
Breeds nearly all year round. Nestlings remain in the nest for

approximately one month after hatching. Young grow to be independent soon after leaving the nest.

### Food

A grain-eater, though largely a scavenger. Being mostly a city-dweller, it relies on garbage and foodscraps in parks, dumps and backyards.

### Emergency food

Adult: budgie mix, crumbed biscuits, bread. Offer plenty of water. Make water available in a dish. Never pour down the throat, as the bird may choke.
Young: grain soaked in water and drained. See FEEDING YOUNG BIRDS page 72.

### How to attract

As these birds are usually found around parks and public places, any food or bread thrown in the air will soon attract their attention. In your yard, they may be attracted to chicken coops or aviaries in search of grain. Erect a bird bath and beware of cats (see page 63 & 64).

### Handling and transport

Pigeons are safe to handle. Their claws are moderately sharp but their bite is not dangerous. See page 64 for instructions on handling birds through a cloth or with bare hands.

See page 65 for instructions on transporting birds.

### Banded pigeons in your yard

There are pigeons, and there are banded pigeons. If a pigeon is banded, it means it is not a feral pigeon, but a pigeon belonging to someone who belongs to a pigeon fanciers' club, or a similar pigeon-oriented organisation. These pigeons are often raced, spending a great deal of time flying from one destination to another. A banded pigeon may end up roosting in your yard. There may be several reasons for this. A change of weather or extreme weather conditions may have forced it to terminate the flight before reaching the destination. It may be exhausted, it may be waiting for weather conditions to improve, or it could be injured. For whatever reason it has chosen your yard as the roosting place, you can be assured it is a temporary arrangement or one brought about by injury. It is not in the bird's nature to change its place of habitat, and instinctively it will fly home upon recovery.

If the bird is not injured, a food supply as suggested for pigeons would sustain it till it has recovered to resume the flight. There are no special housing requirements. Keep your pets away and provide food and water. In days, at the maximum a few weeks, the pigeon will resume its intended flight.

If the pigeon is injured, capture it and take note of the number on the band. Ring the Pigeon Association (See EXTRA HELP page 196); they will be able to track down the owner of the pigeon by the number on the band. If the injury is severe, it may be necessary to take the pigeon to a vet before advising the Pigeon Association of its whereabouts.

Information on the band: four capital letters represent an abbreviation of the club's name. Two digits, for the year of banding. Four digits, owner's code by which, in conjunction with the year of banding, the owner can be identified.

**Did you know?**

*Pigeons made a significant contribution to communication during World War II. As radio messages needed to be sent coded and de-coded upon receipt for security, plain-written messages delivered by pigeons proved quicker in the long run. Only three messages were lost out of approximately 33 000 sent in the southwest area of the Pacific. As a result of a particularly efficient message delivery at Bougainville, two pigeons were awarded the Dickens Medal, otherwise known as the animal Victoria Cross. Those pigeons can still be seen, stuffed, at the Australian War Memorial in Canberra.*

# TURTLEDOVES

## Spotted Turtledove *Streptopelia chinensis*

Other names: Indian Turtledove, Spotted Dove, Indian Dove

Introduced. Active by day.

**Identification**

Average length head to tail 27 to 32 cm. Can be identified by white spots on the back of the neck. Eye: yellow. Bill: grey-black. Legs: red. Young: dull brown all over, without the spotting on the neck; bill: grey.

### Habitat
Lives in pairs. Found in abundance in cities, in parks and suburban gardens. Roosts in trees for the night.
Nest: a platform with a slight depression, made of twigs and sticks. Placed on an outer branch of a bush or a small tree, rarely higher than seven metres above ground.

### Breeding
Breeds at most times of the year. Nestlings remain in the nest for approximately 15 days after hatching.

### Food
A seed-eater. As it is commonly found in cities, it scavenges around parks and yards. It prefers to feed on the ground, looking for seeds of grasses and other plants. Also likes insects and fruit.

### Emergency food
Adults: bird seeds, fruit, crumbed biscuits.
Young: grain soaked in water and drained. See FEEDING YOUNG BIRDS page 72.

### How to attract
As it feeds on a variety of food, it is hard to pinpoint what type of food would attract it to the yard. Birds within this species have different feeding preferences, some for insects, some for seeds. In the yard, it may be attracted to chicken coops or aviaries in search of grain. Erect a bird bath, and beware of cats (see page 63 & 64).

### Handling and transport
Turtledoves are safe to handle as their claws are moderately sharp and their bites are not dangerous. See page 64 for handling birds through a cloth or with bare hands. See page 65 for advice on transport.

## Did you know?
*Though this bird is frequently seen in parks and on the streets, it is not a native to Australia. It was introduced to Australia from India in the last century. It has settled well into the new environment, and is not considered a pest.*

# LIZARDS

Though both are reptiles, lizards differ from snakes in many ways. The most obvious is the presence of limbs in lizards, though on some species, such as the Common Scaly Foot, the limbs are so small they are hardly seen. Those lizards are called legless lizards, and unfortunately, due to their snake-like appearance, they are often killed by humans because of mistaken identity. Some less noticeable differences between snakes and lizards are: most lizards have ear openings, snakes don't; lizards have a solid, fleshy tongue (except goannas, which have a forked one), snakes have a forked tongue; a lizard's lower jaw is firmly joined in the front, snakes' are elasticised for swallowing large prey; lizards have eyelids, some are fixed, others are movable, while snakes have only fixed eyelids.

Many lizards can drop their tails (goannas and shingle-backs are exceptions), snakes can't. If a predator grabs a lizard by the tail, by dropping its tail a lizard may escape. Though a new tail will grow, it could take from three to 12 months, and it will never be as vibrant in colour as the original. The tail is where a lizard stores its fat. If a tail is lost just prior to hibernation, the lizard may starve.

## How to deter lizards

There is no deterrent for lizards. A securely fenced yard, where the fence is dug into the ground and the gate is tight-fitting, will lessen the chances of entry of ground-dwelling lizards. Lizards may enter a yard in pursuit of food and shelter, and the best deterrent is not to create a favourable environment. Keep the yard free of crevices and hiding places, such as corrugated iron sheets, firewood, building materials, or any other piled debris. Keep your yard snail free, particularly if you are bothered by blue-tongue lizards. Uneaten pet food may also appeal to some lizards.

Expert tree climbers, such as goannas, find it easy to gain access to yards in bushy settings. Though it is not easy to deter them from entering a yard, they are not likely to stay long as the food supply is not sufficient to sustain them. Keep your yard free of any food scraps such as barbecue leftovers, as they are sure to tempt a goanna if it is in the area. My leftover barbecued fish were devoured by a visiting goanna, while still in the foil! A dog or a cat in the yard would be some deterrent to lizards. However, dogs cannot be classified as a desirable deterrent, since they are potentially dangerous to much wildlife, including reptiles. If leafy, bushy settings and rockeries appeal to you, it is to your benefit to come to terms with visiting wildlife.

Generally, lizards encountered in urban areas are not aggressive and none are venomous. They do not spread diseases, damage the yard, or make any noise. It is not in their intention to terrorise anyone, and their visits are usually transitory.

## Injured lizard

If the lizard is in the wild, it is probably better to leave it alone and let nature take its course, unless it is suffering and can be captured safely. If the lizard is in your yard or you have caused the injury, use your judgement and assess the severity of the injury. Cosmetic injuries such as a missing toe or a small scratch on the body, do not need attending, as they do not interfere with the animal's ability to hunt for food or run away from predators. Consider the size, as an injured lizard is likely to be aggravated and a large lizard such as a goanna is best left alone. If the lizard is too large to capture safely, report the incident to someone experienced. See EXTRA HELP page 193. Keep your distance and keep your pets away from the lizard. If the lizard can be captured, immobilise its head or claws by wrapping in a cloth. Minor injuries can be treated by applying a cotton ball dipped in iodine or mercurachrome to the wound. Both are available script free from a chemist. For major injuries, take the lizard to a vet. It is common for lizards to have ticks. Unless the infestation is severe and the lizard is in your care, they are best left alone. Note: if injured legless lizards have not been positively identified as lizards, treat as per INJURED SNAKE, page 148. Wash your hands after handling reptiles.

## A lizard in the house

A small skink, or a large goanna, may end up in your house. It is not normal for lizards to enter houses (with the exception of some smaller lizards), and if they do, it is purely by accident and without any bad intentions. They may have wandered in while foraging for food, or looking for shelter. It is in your, and the lizard's, interests to have it vacate the premises. Do not panic, as the lizard will not attack you. If it displays any hostility, it is because it is just as frightened as you are.

Keep your pets away from the lizard. Observe it from a safe distance. If it is near an open door, it may choose to leave. Encourage it to leave by making a clear passage for it. Open the door wide, move any furniture out of the way, close doors or block areas where you do not want it to enter. Having made your intentions obvious, if it still refuses to leave, you can try coaxing it with a long-handled object, such as a broom. If the lizard is small or fragile, be careful not to injure it. Smaller lizards can be cornered and physically removed from the house. Handling procedures are covered on pages relevant to the species.

If the lizard is hostile, large, or you cannot face the situation calmly, consider leaving the house and observing it through a window. Perhaps a neighbour can help. Otherwise, see EXTRA HELP page 193, for a list of people who may be able to help. Herpetologists may advise you further on how to remove it from the house, or may ask you to trap it in the house till they arrive and remove it from your property. Legless lizards in the house should be treated as snakes (see A SNAKE IN THE HOUSE page 145), unless they have been positively identified.

## Euthanasia of lizards

Euthanasia is best and most humanely performed at the hands of a veterinary surgeon. Authorities from National Parks and Wildlife NSW advise that it is illegal for the public to put down native animals. All reptiles are protected species and it is not legal for the public to put them down. However, there may be extreme circumstances which would create an exception. If the reptile is suffering, the condition is irreparable, and help is not available, it may be more humane to kill the reptile on the spot. Provided you have access to a freezer, freezing is the most humane method for reptiles. Place the reptile in a bag, and place in the freezer. A couple of hours should kill most species, including larger reptiles such as goannas. Remove from the freezer and dispose by burying.

Should you have no access to a freezer, use your commonsense. Small reptiles can be squashed with your foot, or by dropping a heavy object onto their body. Larger lizards can be run over by a car, or hit over the head with a sharp object such as a spade. Before proceeding with euthanasia, give yourself one last chance to assess the situation. Is euthanasia the only solution? Can the reptile still be saved at the hands of an expert? Is there no vet anywhere within reach? Could someone else deliver the reptile to a vet? Having decided on euthanasia, it is your moral obligation to ensure the job is done with minimal pain to the reptile, and to the point when the reptile is dead. Wash your hands after handling reptiles.

Important note: These procedures have been written in the humane interest of minimising the suffering of reptiles.

# BLUE-TONGUED LIZARDS

## Eastern Blue-Tongued Lizard *Tiliqua scincoides*

Native. Active by day. Not venomous.

### Identification

Average length — 30 to 45 cm. Body is quite stocky, with small limbs, a short rapidly-tapering tail, and a distinct blue tongue. Body colour varies from grey to brown, and body is crossed with dark bands from the neck to the tail tip.

### Habitat

Leads a solitary life. Found throughout Australia in a variety of suitable conditions. Likes to shelter in hollow logs and under debris. For that reason, it is often found in backyards, under piled rubbish, such as corrugated iron sheets, in bushes and among timber.

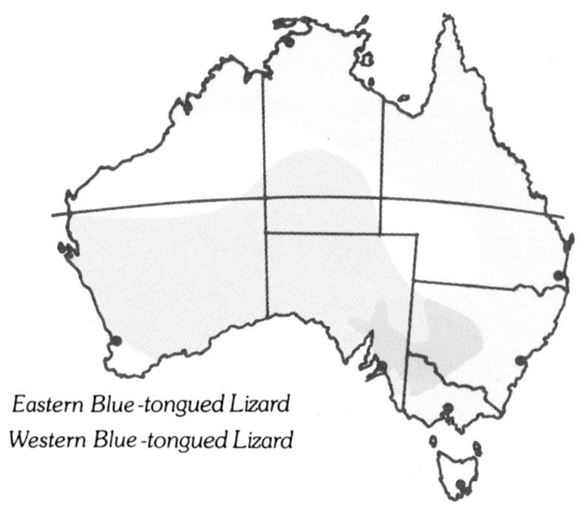

**Breeding**
Bears live young in summer and early autumn. Litters range from six to twenty. Young average 10 cm in length and are similar to the parents, but with more pronounced markings and in a more vivid colour. The young fend for themselves from birth.

**Food**
Will consume both vegetable and animal matter. Feeds on snails, carrion, fruit, flowers and berries. In a garden, may eat baited snails or snail baits, which may kill it.

**Emergency food**
Snails; canned dog or cat food; chopped raw meat; soft fruit; raw eggs. In an emergency, can survive without food for at least a week, especially during cold weather.

**Hostile behaviour**
When alarmed, positions itself in a curve, resembling letter 'C', always facing the opponent. Flattens its body, sticks the tongue out and hisses loudly.

**How to attract**
There is no way to specifically attract a blue-tongued lizard; however, creating a suitable environment in your yard would help, particularly if you live near the bush. Lots of suitable sheltering places, such as hollow logs, rockeries, low shrubs, and a good population of snails, would certainly be appealing should a lizard be in your area.

However, other reptiles may also find such habitat attractive. It is therefore a good idea to create such an environment away from your house.

## Handling

Should it be necessary for you to handle this lizard, keep in mind that all wild animals resent being handled and may attempt to bite. If injured, the lizard may be more aggressive, though some injuries may make it less mobile. Always have a container/bag ready prior to capturing the lizard. Handling blue-tongued and most other lizards is safe, provided you stay clear of the mouth, and to a lesser degree the claws. For extra safety and confidence, the following procedure can be attempted wearing gloves, or by first placing a cloth or towel over the lizard. With one hand, grab the lizard behind the neck, enclosing its body and front limbs in your hand, ensuring that the limbs point towards the tail. Lift the lizard. For larger specimens, lift the lizard's body with your other hand, with the hind limbs also pointing towards the tail. Avoid putting too much pressure on the tail. Wash your hands after handling lizards. Blue-tongued lizards' bites are not venomous and most bites result in superficial wounds only. See FIRST AID, LIZARDS for more information page 11.

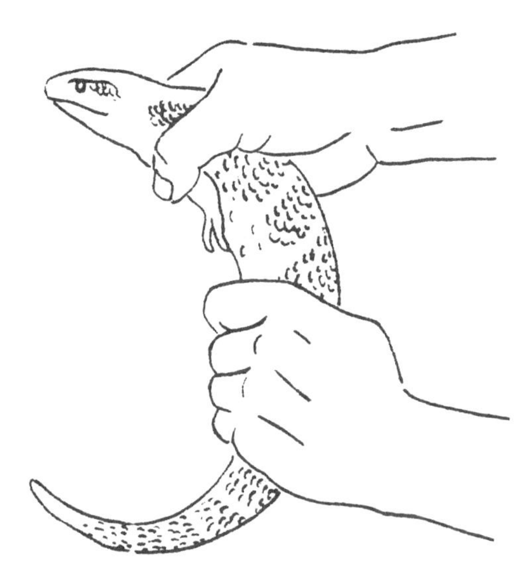

## Transport

Lizards can be transported in light-coloured, strong bags, tied at the top to stop the lizard escaping. A pillow case is a good alternative, provided there is no weakening at the seams. Turn inside out, to stop the lizard from entangling in the seams. Tie the top with string to prevent the lizard from escaping. Cardboard or shoe boxes are also suitable for transporting. Line with grass, punch holes for air, and ensure the lid fits tightly or cover securely. Reptiles may suffer from overheating, and should not be kept in warmth for long periods. Sprinkle some water onto the grass for a lengthy journey, as lizards, particularly small ones, dehydrate quickly.

### Did you know?

*When aggravated, this lizard puffs out and curves its body. That makes it look like a death adder, a poisonous snake. Many a time, a herpetologist has travelled in vain to remove a 'dangerous snake' from a distressed caller's yard, to be confronted by a Blue-tongued Lizard.*

## Shingle-back Lizard *Trachydosaurus rugosus*

Native. Active by day. Not venomous.

### Identification
Average length 35 cm. Body is quite stocky with short limbs and the head and tail tapering alike. Colour varies regionally. Top colour can be any shade of brown, from reddish to almost black. Some specimens are scattered with cream or yellow spots or bands across the body. Belly is whitish. Scales are large and raised, giving the lizard the appearance of a pine cone.

### Habitat
Leads a solitary life. Common in dry inland areas, inhabiting heavily timbered areas as well as open grasslands. Shelters under fallen timber, leaf litter and under shrubs and grasses.

### Breeding
Bears live young. Litters are small, one, two, rarely three offsprings. Young measure about half the size of adults, and fend for themselves from birth.

### Food
Basically a vegetarian, though also eats insects and snails. Main foods include ground vegetation such as

blossoms, herbage, fruit, and berries. Fat is stored in the tail for nourishment during hibernation.

**Emergency food**
Ripe fruit such as bananas; snails; beetles; raw meat; raw eggs. In an emergency, can survive without food for at least a week, especially during cold weather.

**Hostile behaviour**
This slow-moving lizard is inoffensive and prefers to retreat, though it will bite if provoked. When cornered, it opens its mouth, darting its tongue and hissing loudly.

**Handling**
Handled in the same manner as blue-tongues. See EASTERN BLUE-TONGUED LIZARD, HANDLING page 126.

**Transport**
Most lizards can be transported in a similar fashion. See EASTERN BLUE-TONGUED LIZARD, TRANSPORT page 126.

# DRAGONS

## Bearded Dragon *Amphibolurus barbatus*

Native. Active by day. Not venomous.

## Identification
Average length is 40-50 cm, though in some areas may reach 60 cm. Top of body is uniform in colour, ranging from shades of grey to shades of brown, always matching the environment for camouflage. Belly is whitish. The inside mouth is bright yellow. Throat has a well developed pouch or 'beard', with spines across, to the top of the head. This dragon is often confused with the Frill-necked Lizard. The frill of the Frill-necked Lizard extends around the head, unlike the 'beard' of the dragon, which stops at the throat.

## Habitat
Leads a solitary life. These lizards are comfortable both on land and in trees. They can often be seen basking in the sun on fallen timber, tree stumps, on fence posts, and on road verges.

## Breeding
Female deposits from eight to 25 eggs in an excavated hole, and buries them for incubation and protection. Young resemble the parents, but have more vivid colours and a less pronounced beard. They fend for themselves from hatching.

## Food
Insects are the main diet, though also feeds on flowers and soft herbage.

## Emergency food
Insects such as crickets, cockroaches, and moths; chopped lean meat; chopped fish; snails; bananas; chopped lettuce. In an emergency, can survive without food for at least a week, especially during cool weather.

## Hostile behaviour
The first sign of distress is a change in body colour. The head, tail, and limbs become darker. Loose skin around the neck erects, and the mouth opens exposing the bright yellow interior. Further provocation may result in hissing and flattening of the body.

## Handling
Handle in the same way as blue-tongued lizards. See EASTERN BLUE-TONGUED LIZARD, HANDLING page 126.

## Transport
Most lizards can be transported in a similar fashion. See EASTERN BLUE-TONGUED LIZARD, TRANSPORT page 126.

## Did you know?
*It's not only when this lizard is under stress that its body changes colour. This also occurs during sexual activity, injury or a variation in temperature.*

# GOANNAS OR MONITORS

## Handling

Handling these lizards is not recommended, except by qualified people. In an emergency, should it be necessary for you to capture a goanna, remember all wild animals resent being handled and may attempt to bite. If injured, they may be more aggressive, though some injuries may make them less mobile. Goannas are potentially dangerous as they have sharp, backward-pointing teeth, sharp claws, and a strong tail. Handle wearing heavy duty gloves, or having first placed a thick towel or blanket over the lizard. Goannas can be handled by quickly grabbing the back of the neck with one hand, and the hind limbs and base of tail with the other. Hold the lizard so its claws are facing away from your body. Another way to handle is by the tail. Quickly grab the goanna by the thick part of the tail, and lift. Hold away from your body, especially your legs, with your arm stretched out. Swing the lizard gently, to stop it from climbing up your body and biting you. Larger goannas are best captured by slipping a noose around their neck. The noose can be made from a bamboo rod and

fishing line or fine wire. If the goanna needs transporting, have the transport bag ready prior to capture. Wash your hands after handling reptiles. For information on how to treat a goanna bite see FIRST AID, LIZARDS, page 11.

### Transport

Transport in a light-coloured bag made of material strong enough to withstand goannas' sharp claws. The size should be large enough to slip the goanna in easily. The corners should be sewn off to prevent the lizard from getting entangled. Tie the top firmly with string, to stop the lizard escaping. Reptiles may suffer from overheating, and should not be kept in warmth for long periods.

### A goanna in the yard

While foraging for food, a goanna may end up in your yard. Do not be alarmed, as one yard could not sustain a goanna for long, and it is either passing through, or has been attracted to some food. Keep your pets away. Observe the goanna from a safe distance, as tracking its movements will satisfy you that it has truly left the premises. From experience, one tends to tire of this chore, and when you look up at some point, it has mysteriously disappeared. Do not attempt to chase or feed it. It will move when it is ready, probably when it thinks you have stopped observing it. Should it become necessary to have it removed, handling procedures are covered on page 130. However, goannas are best handled by experienced people, and EXTRA HELP page 193 lists people who may be able to oblige. If you see the goanna heading for your neighbour's yard, a courtesy call may be appreciated. If nothing else, a goanna in a yard makes good conversation at parties.

### Did you know?

*Some goannas incubate their eggs in an unusual way. The female digs a hole in a termite nest, lays her eggs, and leaves. The termites quickly go about mending the damage, sealing the hole and in the process, the eggs. In the warmth of the nest, away from predators, the eggs hatch.*

### Did you know?

*The name 'goanna' probably came into existence when the early settlers mispronounced the name 'iguana', a similar lizard from the American continent. Goannas are the only Australian lizards with forked tongues. As in snakes, it is not the nostrils but the tongue which is instrumental in its sense of smell.*

## Tree Goanna *Varanus varius*

Other names: Lace Monitor

Native. Active by day. Not venomous.

### Identification

Average length is 1.5 metres, though may exceed two metres. Has a slender body, with strong limbs and a long neck. Body colour is blue-black, with yellow spots forming bands across the body. Belly is cream.

### Habitat
Leads a solitary life. Prefers dry country. Lives mainly on the ground, though forages on both trees and the ground. Shelters in hollow logs or under bark. When alarmed, climbs a tree with great ease.

### Breeding
Female deposits from six to 12 eggs in a warm protected spot such as a termite mound, in a hollow log, or in a litter below the surface. Young average 28 cm in length, and are similar to the parents but with more pronounced markings and in a more vivid colour. They fend for themselves from hatching.

### Food
Carnivorous. Feeds on birds, eggs, mammals such as rats and mice, reptiles, insects, and carrion. Has been known to take scraps from public places such as parks.

### Emergency food
Raw meat; canned dog or cat food; raw eggs; fish. In an emergency, can survive without food for at least a week, especially during cool weather.

### Hostile behaviour
Generally not an aggressive creature, and will choose to retreat, running up the trunk of the nearest tree, keeping to the side away from the intruder. If cornered, will inflate the body, hiss loudly, and lash with the tail, in a bluffing tactic.

### Handling and transport
See GOANNAS, HANDLING and TRANSPORT page 130 & 131.

## Gould's Goanna *Varanus gouldii*

Other names: Sand Monitor

Native. Active by day. Not venomous.

### Identification
Average length is one to 1.5 metres, though it has been known to exceed that. Has a slender body, with strong limbs and a long neck. Colour varies regionally, from yellowish to almost black above, and is marked with yellow markings, forming cross-bands or irregular patterns, across the body. Belly is usually cream, speckled with dark spots.

### Habitat
Leads a solitary life. Basically a ground dweller, though will climb a tree if cornered. In forests, shelters in hollow logs or under tree debris. In dry country, digs burrows for shelter or enters disused rabbit warrens.

### Breeding
Female deposits from five to eight eggs in a warm, protected spot such as a termite mound, hollow log or in litter below ground surface. Young average 28 cm in length, and are similar to the parents but more vivid in colour and with more pronounced markings. They fend for themselves from hatching.

### Food
Carnivorous. Feeds on birds, mammals such as mice, rats, and rabbits, insects, reptiles, and carrion.

### Emergency food
Raw meat; canned dog or cat food; raw eggs; fish. In an emergency, can survive without food for at least a week, especially during cool weather.

### Hostile behaviour
Chooses to retreat, running off sometimes on two hind legs, or running up the trunk of the nearest tree. If cornered, will inflate the body, hiss loudly, and lash with the tail, in a bluff.

### Handling and transport
See TREE GOANNA, HANDLING and TRANSPORT page 130 & 131.

# GECKOS

## Southern Leaf-tailed Gecko *Phyllurus platurus*

Other names: Broad-tailed Gecko

Native. Nocturnal. Not venomous.

### Identification

Average length 15 cm. Body gives the impression that it is covered with prickles. Head and tail are similar, triangular shaped. Body colour is a mixture of shades of brown, salt and pepper in appearance. Belly is cream or white.

### Habitat

Basically solitary, though groups can be found in favourable habitats. Inhabits sandstone areas, sheltering under rocks, in rock crevices, and in small caves in sandstone. Occasionally found in garages and in houses.

### Breeding

Female deposits an average of two eggs, rarely one, in a secluded spot such as under bark of trees, logs or stones. Young fend for themselves from hatching.

## Food
Insectivorous. Feeds on insects, worms, invertebrates and mealworms.

## Emergency food
A variety of insects such as crickets, cockroaches, moths; mealworms. In an emergency, can survive without food for up to a week, especially during cool weather.

## Hostile behaviour
Although not hostile, when hurt or feeling threatened, makes a squealing sound resembling a miaowing cat, opens its mouth wide and wriggles its tail.

## How to attract
They cannot be attracted by any specific means. Being fond of sandstone, they are more likely to enter dwellings made of bricks or concrete. An old wooden cupboard against a brick wall, and an undetected termites' nest in my garage, proved an excellent dwelling for my inhabitants.

## Geckos in the house
Do not be alarmed at the sight of a gecko in your house. This docile creature will not attempt to move away, even upon close inspection. Geckos are not dangerous or troublesome. They don't chew, make a noise, or spread disease. Being nocturnal, they are seldom seen, evidence of their presence being small, odour-free, dry droppings. Consider sharing your house with them. In return, they will keep down the population of some pest insects. Removing one gecko from your house may not solve your problem, as the rest of the family is probably somewhere else on your property. Do not be concerned about being overrun by geckos, as they are not prolific breeders, usually producing only two offspring.

## Handling
Have a holding container ready prior to capturing the lizard. This lizard is safe to handle, provided you stay clear of its mouth, as it may attempt to bite when on the defensive. Though geckos have teeth and are capable of biting, their bites are harmless and not infectious. If bitten by a gecko, see FIRST AID, LIZARDS page 11. Handle with lightweight gloves, if you wish.

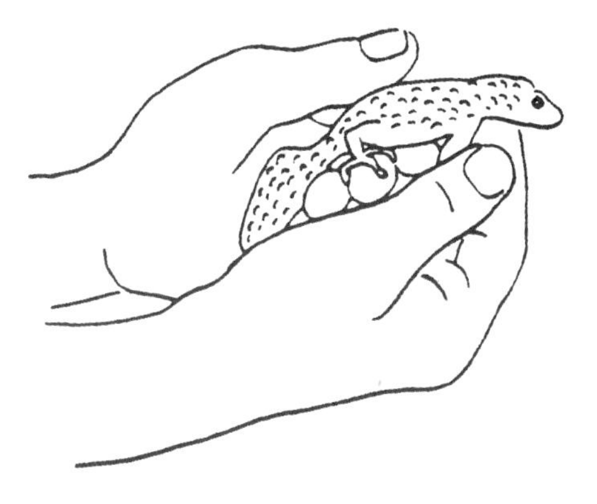

Do not wear heavy duty ones, as you may injure the gecko. Place your hand over the lizard, immobilise its head between your thumb and middle finger, and lift. Once lifted, it can rest in the palm of your other hand. Do not put any pressure on the tail. Another method of handling is to sweep onto a dustpan or a cloth, and lower into the transport/holding container. Wash your hands after handling animals.

## Transport
Transport in a shoe box, an ice-cream container or any cardboard box. Line the bottom with grass,

punch holes for air and ensure the lid is tight fitting, or cover securely. Being small, it dehydrates quicker than larger reptiles, so sprinkle water onto the grass, for lengthy keeping. Reptiles may suffer from overheating and should not be kept in warmth for long periods.

# LEGLESS LIZARDS

## Common Scaly-foot Lizard *Pygopus lepidopodus*

Native. Nocturnal, sometimes active by day. Not venomous.

**Identification**
The largest of all legless lizards. Averages 60 cm, though may reach 75 cm. Colour and patterns vary, but generally the top is a shade of grey, brown or red-brown, marked with dark blotches or stripes. Belly is a shade of cream. Though it may be mistaken for a snake, it is a legless lizard.

**Habitat**
Leads a solitary life. Found throughout Australia in many forms of habitat, from wet to dry areas. Shelters under rocks, timber and leaf litter.

**Breeding**
Female deposits two, sometimes three, eggs in a secluded spot, such as under debris, logs or rocks. Young resemble the parents, with more vivid colours and more pronounced markings. They fend for themselves from hatching.

**Food**
Insectivorous with a liking for spiders. Foods include spiders such as trapdoors, a variety of insects and small lizards.

**Emergency food**
Small insects such as beetles, grasshoppers, spiders, and may take soft fruit. In an emergency, can last without food for up to a week, especially during cool weather.

**Hostile behaviour**
Although shy and quick to retreat, will display open aggression if provoked. Then, it will lift its head and neck, forming an 'S' shape, and strike like a snake.

**Handling**
If it has not been positively identified as a lizard, handling is not recommended except by qualified people. In an emergency, should it be necessary for you to capture the lizard, remember, all wild creatures resent being handled and may bite when on the defensive. If injured, they may be more aggressive, though some injuries may make them less mobile. Though not venomous, these lizards are capable of inflicting a painful bite. Bites, in most cases, would not draw blood. If the skin breaks, see FIRST AID, LIZARDS page 11.

Have a transport/holding container ready prior to capturing the lizard. Correct handling relies on immobilising the head to stop the lizard from biting. It is easier to grasp if the head can be pinned down first with some long-handled object, such as a golf club. With one hand, grasp the lizard by the back of the neck, immobilising its head between your fingers — index finger on top of the head, thumb and middle finger either side of the head. With the other hand, lift the lizard's body. This lizard can drop

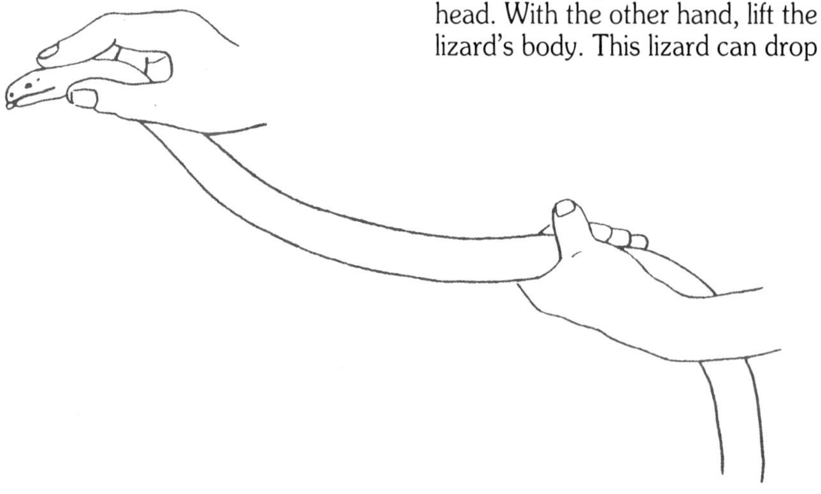

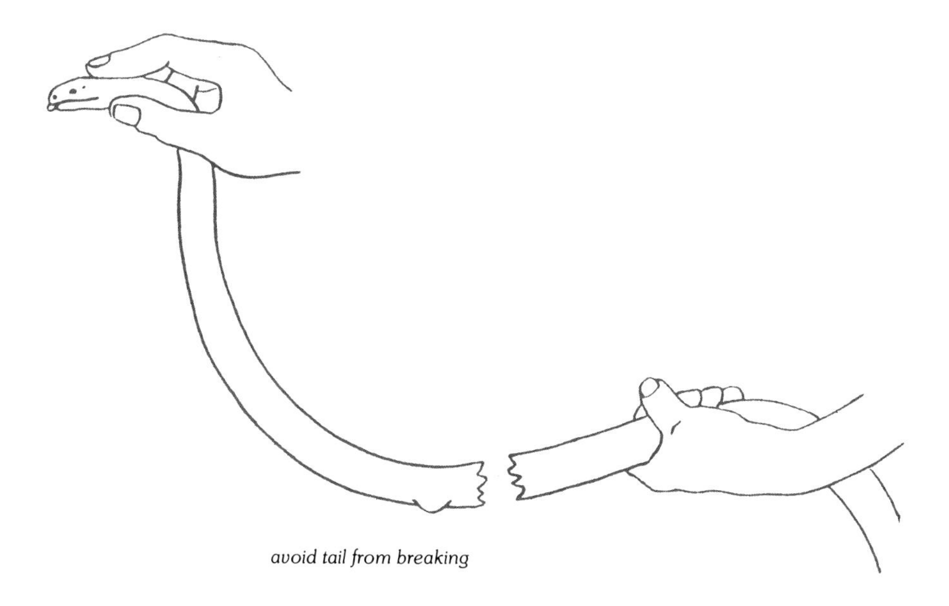

*avoid tail from breaking*

its tail, which is 60 per cent of total body length. Therefore, handling should be confined to the 40 per cent of the area starting at the head. Wash your hands after handling reptiles.

**Transport**

If the lizard has been positively identified, it can be transported in a bag. A strong, light-coloured pillow case is a good alternative. Turn inside out to stop the lizard from entangling itself in the seams. Tie the top with string, to stop the lizard from escaping. It can also be transported in a cardboard box. Ensure it has air-holes, and a tight fitting lid, or is securely covered. Line the bottom with grass. Being a small species, it dehydrates more quickly than larger reptiles. Sprinkle some water onto the grass for lengthy keeping. Reptiles may suffer from overheating, and should not be kept in the warmth for long periods. Note: If the species has not been positively identified as a legless lizard, transporting should be left to experienced people.

**Did you know?**

*During evolution, a shortening of the limbs has taken place in some reptiles, of which the Common Scaly-foot Lizard is one example. The most extreme example of this process are snakes. Theoretically, at one time in the past, they, too, may have had limbs. What makes the Common Scaly-foot a lizard is the fleshy tongue, the ability to drop the tail, and ear openings, among other things.*

# TORTOISES

A tortoise or a turtle? There is some confusion about the correct term for these docile creatures. In Australia, the large marine species with the flippers are known as turtles. The smaller ones with claws, ones inhabiting rivers and lakes, are called tortoises.

There are still unanswered questions about the habits of tortoises. Though they spend most of their time in the water, emerging to bask or lay their eggs, why do they leave the area and journey overland? Large numbers of tortoises will sometimes migrate, for reasons which are not fully understood.

## A wandering tortoise

Should you see a tortoise wander into your yard, or be wandering on the road, do not be afraid to pick it up. See HANDLING, TORTOISES page 142. Examine it for injuries, especially the shell. If the tortoise is injured or the shell is cracked, take it to a vet. If the tortoise appears well, it can be released in a suitable area. Tortoises must have fresh or brackish water for survival, as per TORTOISES, HABITAT page 142. Find such a spot, preferably away from traffic and people. Do not release in a national park without prior consent from your State Fauna Authority. See EXTRA HELP page 193. They may also be able to advise you on a suitable release spot. Wash your hands after handling tortoises.

## Injured tortoise

If you see an injured tortoise in the wild, it may be difficult to capture, therefore it is best left for nature to take its course. Some injuries are cosmetic, such as a missing toe or a small scratch. If they do not interfere with the animal's ability to get away from predators or hunt for food, they are best left alone. If the tortoise is in your care, or you have caused the injury and captured the animal, minor superficial wounds can be treated. Put a few drops of iodine or mercurochrome onto a cotton ball, and dab onto the wound. Both are available script free from chemists. Should you feel threatened by the tortoise, the head or limbs can first be immobilised in a cloth. If the injury is major or the shell is cracked, take the tortoise to a vet. Wash your hands after handling tortoises.

### Did you know?

*Ideally, tortoises must have both land and water for survival, therefore they cannot live in aquariums without a basking area. Unlike some exotic species, Australian tortoises do not have a tongue. Therefore they consume food in water. Food is placed in the mouth, and the head submerged in water. With the help of neck muscles, the water carries the food down the throat, and into the stomach. Afterwards the water is expelled.*

# Long-necked Tortoise *Chelodina longicollis*

Other names: Eastern Snake-necked Turtle

Native. Active by day. Not venomous.

**Identification**
Average shell length is 20 to 25 cm. Shell is mostly oval, though can be egg-or pear-shaped, and is uniform in colour. Top varies from light brown to almost black. Bottom is cream, and underparts edged in black. Tail is extremely short, neck long, almost half the length of the shell.

**Habitat**
Leads a solitary life. Must be near water, either fresh or brackish. Prefers quiet waters, such as lakes or swamps, to fast-flowing rivers. Can bury itself under debris or leaf matter, to rest or for hibernation.

**Breeding**
Breeding usually takes place during summer. Female lays up to ten eggs (though up to 24 have been recorded), in a hole excavated near the water which it occupies. Eggs hatch in three to four months. Young fend for themselves from hatching.

**Food**
Basically carnivorous, but will eat a small quantity of plant matter, such as waterlilies. Basic diet consists of small river crustaceans, water snails, worms, small fish, and tadpoles.

**Emergency food**
Food must be offered near water, as it is consumed in the water. Offer lean mince; chopped fish or whole small fish; canned dog or cat food. In an emergency, can survive without food for at least a week, especially during cool weather.

**Hostile behaviour**
Does not display any hostile behaviour. Has been known to become snappy when fed in captivity.

**Handling**
Handling these tortoises is safe. Stay clear of the mouth as like all animals on the defensive, it may attempt to bite. The claws are moderately sharp. Have a transport/holding container ready, prior to capturing. Gloves can be worn for extra safety, or the tortoise can be picked up through cloth or a small towel. Pick up with one hand, by the back of the shell or, with two hands, lift by the sides of the shell. Wash your hands after handling tortoises.

**Transport**
Transport in a cardboard or shoe box. Line the box with grass, punch holes for air, and cover securely with a lid to stop the tortoise escaping. Reptiles may suffer from overheating, and should not be kept in the warmth for long periods.

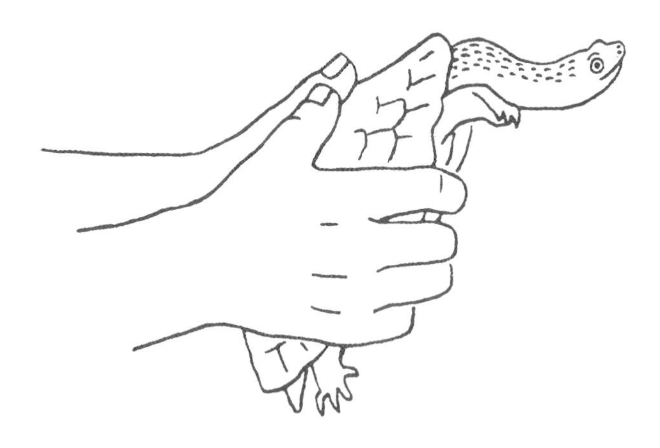

# SNAKES

Most people think of snakes as slimy, slithering and generally undesirable. Yet they are not slimy, nor even wet. Their skin is quite dry. Like all reptiles, snakes shed their skin. It is the result of growth, and occurs more frequently when there is an abundance of food, such as in captivity. A snake will shed its skin several times a year. Prior to shedding, its eyes turn blue, temporarily blurring its vision, until the day of shedding. The snake rubs itself on a rough surface, such as a branch, until the skin splits at the mouth. It then crawls out of its skin, leaving it inside out. Larger snakes such as pythons, cast off their skin in several pieces.

Though snakes have nostrils, they are not used for smell, but for breathing. Smell is activated by the flickering of the tongue. Minute particles of scent are picked up when the tongue is out—and delivered to the roof of the mouth. There, they are activated by a sensory receptor known as Jackobson's organ.

## How to deter snakes

There is no effective way of deterring a snake. If your property is completely fenced off with a brick or metal fence dug into the ground, and with a tight gate fitted with a rubber flap at the bottom, you minimise your chances of getting snakes. The best deterrent is to concentrate on not creating a favourable environment for their basic needs, food and shelter. Keep your yard tidy and free of junk, such as corrugated iron sheets, logs, timber off-cuts, building materials, and firewood. If storage of any of the above is necessary, store them on an elevated platform, about half a metre above ground.

Keep your grass cut short, and any vacant allotments nearby clean. A farm, an aviary, or a chicken coop may attract mice, which in turn may attract snakes. Store fodder in mice-proof bins, and keep the area clean and mice free.

Snakes such as red-bellied blacks, which are fond of damp places, may be attracted to pool areas. Ensure that soil around above-ground pools is packed well near the base. Keep ground area near in-ground pools on sloping blocks as free of crevices as possible. A population of kookaburras on your property may help keep down the snake population; however, they are most effective against snakes up to 30 cm long. A dog in the yard may discourage a snake from entering.

## Handling

Handling of venomous or unidentified snakes is not recommended, except by qualified people. Should it be necessary to capture the snake, see EXTRA HELP page 193 for a list of people who may be able to help. In an emergency, if you must capture the snake, remember all wild animals resent being handled and may attempt to bite. If injured, they may be more aggressive though some injuries may make them less mobile. Have a container/sack ready prior to capturing the snake. Correct snake handling relies on immobilising the head, to stop it from biting. The following procedure is one recognised by most snake handlers. The snake's head is pinned down by a long-handled object called a jigger. Holding onto the jigger with one hand, the other hand grasps the snake's head from the top, by the back of the neck, immobilising the head between the fingers — index finger on the top of the head, thumb and middle finger either side of the head. (Snakes with large heads, such as pythons, require the whole hand to wrap around the snake's head.) The snake is lifted by the neck, keeping it immobilised at all times. The other hand is used

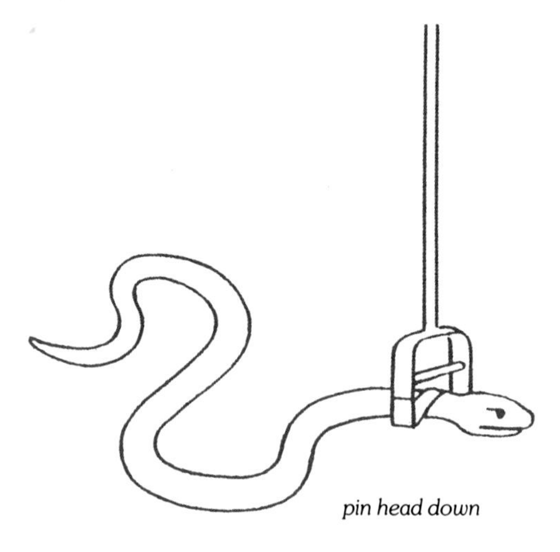


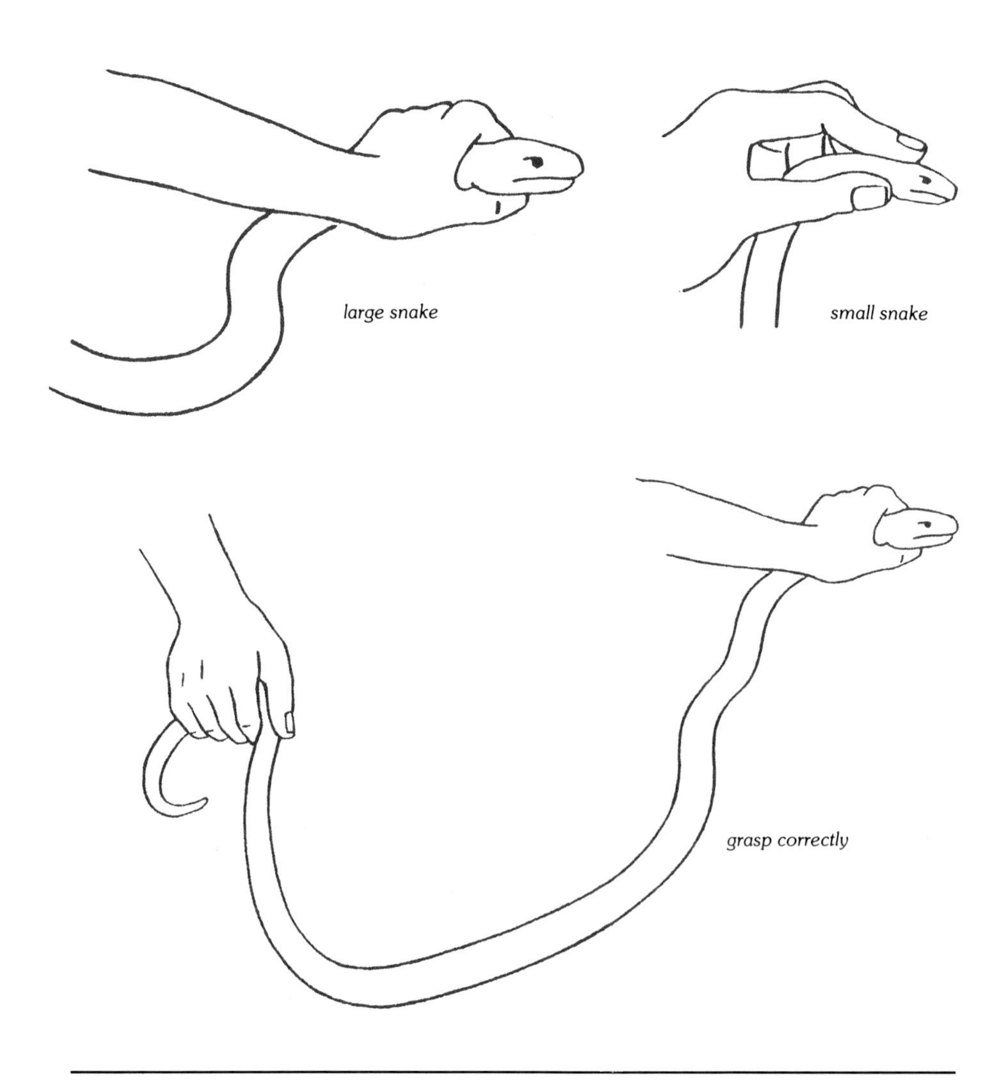

to lift the snake's body, and to stop the snake from coiling around the handler. (Small snakes can be lifted by the neck, using only one hand.) The snake is then lowered into a bag, tail first. This procedure is effective when the snake is on a fairly flat surface, and may vary to suit other situations. Snakes can also be picked up by the tail, or levered by a stick halfway down the body. As the head is not immobilised, these methods require more skill to avoid being bitten. Wash your hands after handling snakes.

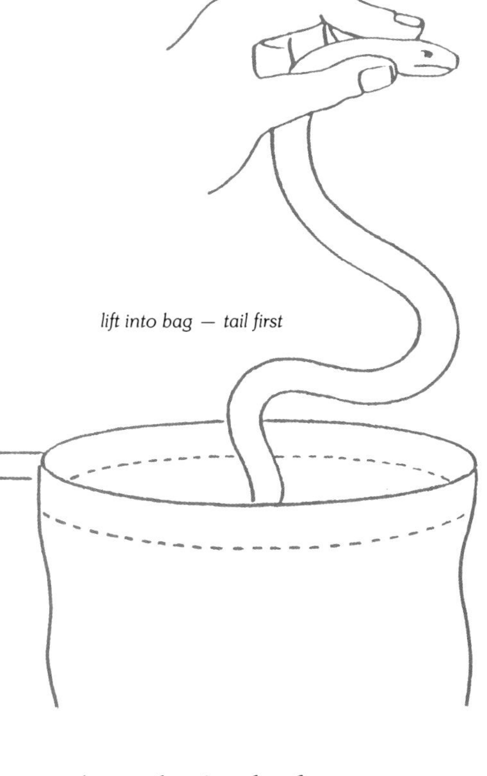


## Transport

Transporting venomous snakes should be left to experienced people. They use light-coloured bags, made from strong material, tied at the top to stop the snake escaping. A light-coloured strong pillow case is a good alternative, provided there is no weakening of the seams. Turn inside out to stop the snake entangling in the seams. Reptiles may suffer from overheating and should not be kept in warmth for too long.

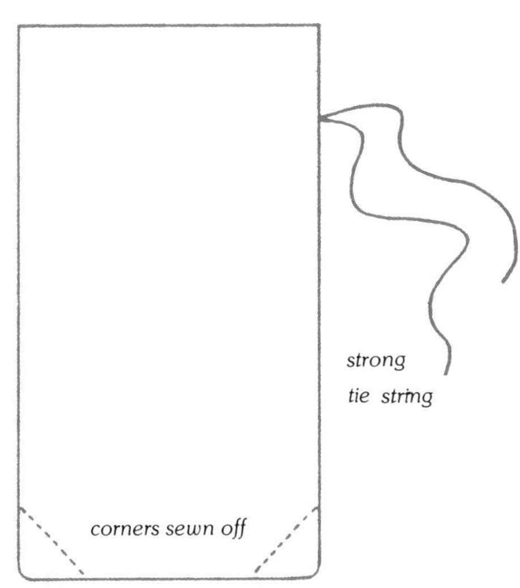


## A snake in the house

A snake may end up in your house while looking for food or shelter. Not all snakes are venomous, but unless you're experienced in identifying snakes, it is better to assume that you are dealing with a potentially dangerous creature. Do not panic, reserve your energy for a plan of action. Keep your pets and any inquisitive children away. For your piece of mind, keep track of the snake's movements.

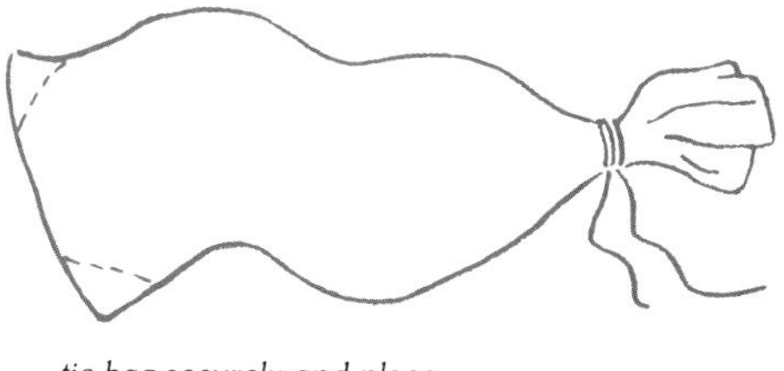

*tie bag securely and place in a spot which is not too warm*

Observe it from a safe distance, if necessary through a window. If it is near an open door, it may choose to leave. If possible, encourage it to leave by making exit accessible. Open doors for it to leave, or close doors to areas which you do not wish it to enter. Do not take unnecessary chances; some doors can be pushed open from a distance using a broomstick, or leave the house through a window and open the door from the outside.

Having provided the snake with a path to leave, allow it some time in which to vacate. Leave the room and observe it; it should eventually leave. If you have not succeeded in encouraging the snake to leave, if you cannot deal with the situation, or if the snake poses a threat, see EXTRA HELP page 193 for a list of people who may be able to help. They may ask for the snake's size, colour, and if the head is distinct from the neck. They may advise you on further procedures to remove the snake, or may ask you to trap it in the house to be certain of its removal upon their arrival.

In either case, unless you see the snake physically removed off your property, or see it voluntarily leave, you may not feel at peace. It is for that reason that you should keep an eye on the snake's movements till the situation is resolved. Having solved your problem, should you see the snake wandering into your neighbour's yard, a courtesy call would probably be appreciated.

## A snake in the yard

You may see a snake in your yard. If it is lying calmly in a sunny spot, it is sunning itself and should shift once it has reached desired temperature or when the sun goes down. It may be just passing through your yard in its quest for food. Keep calm, spend a moment observing the snake. It is a good idea to take note of its approximate size, colour and if the head is distinct from the neck. That will help identify the snake should you call for help. Keep your pets away, and do not poke or provoke the snake. Many herpetologists have had the unnecessarily difficult task of capturing a snake which has grown hostile through aggravation. Your best solution is to wait for the snake to move when it is ready.

If you feel threatened and want the snake removed, see EXTRA HELP page 193 for a list of people who may be able to help. If two people are present, one should observe the snake while the other makes the call. If alone, do the best you can to keep track of the snake while you place the call. For peace of mind, it is better to see the snake leave, or see it removed from the property. Should the snake be heading for your neighbour's yard, a courtesy call would be appreciated.

## A snake in the pool

It is a reasonable assumption that snakes in pools are there by accident. If the pool is full to the top, the snake may be able to get out. (See SNAKE IN THE YARD page 146, if confronted by the snake.) If it cannot get out, it will swim frantically, slowly exhausting itself. At that point, it should sink to the bottom. Stamina varies within species; some snakes could swim for many hours before falling to the bottom, while others tire more quickly. A snake at the bottom of a pool is not a sign that it is dead, though eventually it will die underwater. The snake may end up in the filter basket, where it may remain trapped, but not necessarily dead.

Though not all snakes are poisonous, it is safer to presume that you are dealing with a potentially dangerous snake, and not to take any chances. Observe the snake, taking note of its colour, approximate length, and if its head is distinct from the neck. Unless you are confident about handling snakes, call for help. See EXTRA HELP page 193 for a list of people who may be able to come to your rescue. They may advise you on how to remove the snake, or may come out to remove the snake themselves. Either way, being able to describe the snake will be useful in its identification. If they identify the snake as not venomous, you may be advised to scoop it out with your pool scoop, and release it in bushland nearby.

## A snake in the bush

If you come across a snake while strolling through the bush, you should respect the fact that you are now in its rightful territory, and at best, share it with it. It is not in the snake's interest to bite you, therefore it should not be your aim to injure, capture, or kill it. Reptiles are protected by law, and cannot be killed unless they are a threat to life, stock or property. Do not confuse basking in the sun, or foraging for food, with a threat. If you aggravate the snake, corner it or appear to be a threat, you risk being bitten. The best solution is to move away from the area where the snake is, walking off at normal pace, if the snake does not move first.

Generally, snakes do not attack people. Most bites are the result of treading on a snake, handling it incorrectly, or possibly cornering and teasing the snake. It is also not in snakes' nature to chase after people. If you wish to observe the snake, you may do so at a safe distance: a radius of three metres has been recommended by herpetologists. Most snakes can strike to a distance of a third of their body length, though there are exceptions where they can strike to their own body length. If you wish to identify the snake later, take note of its colour, approximate length, and if the head is distinct from the neck. If your dog is with you, keep it away from the snake.

## A snake in a tent

A snake in a tent has probably wandered in while foraging for food or looking for shelter. Do not panic, and do not throw anything at the snake. Reserve your energy for planning your next move. Unless you can positively identify the species, it is better to assume you are dealing with a potentially dangerous snake, and not to take any chances.

First, try giving the snake an opportunity to leave of its own accord. Open the tent flap wide, and observe from a distance of about three metres. Give the snake a couple of hours, or till the sun goes down. If it hasn't shifted, try stamping your feet at the back of the tent, on the opposite side to the entrance. The vibration may cause it to leave. Or, away from the entrance, try prodding the tent floor from underneath, with a stick. Lift the floor area near the snake, levering it towards the exit flap. I have been assured that no snake would still refuse to leave after such treatment.

Failing that, the only solution is to book yourself into a comfortable motel nearby.

## Pet bitten by a snake

If your pet has been bitten by a venomous snake, any of the following symptoms will appear soon after: paralysis of legs or whole body; salivating; increased heart rate; laboured breathing; vomiting; dilated pupils. Take the animal to the vet immediately. These symptoms also apply to tick bites; however either condition requires a trip to the vet. Symptoms from non-venomous snakes, such as pythons, would be small puncture marks which may bleed slightly. No treatment is necessary. Should the bite area become swollen or pussy, it may be a sign of infection. Take the animal to a vet.

## Injured snake

An injured snake is a dangerous snake. Do not go anywhere near it, and observe it from a safe distance of at least three metres. If it is in the bush, unless you're accompanied by experienced snake handlers, leave it alone and let nature take its course. If it is in your yard, apply procedures for A SNAKE IN THE YARD, page 146. Be extra cautious, and keep your pets and children away. If you have a dog, examine it for possible puncture marks and snake bite symptoms in case it is responsible for the snake's injuries.

## Euthanasia

When confronted by any snake, the possible risk to human life should take priority. When confronted by a severely injured unidentified snake, unless you are experienced with snakes, leave it and let nature take its course. Handling by inexperienced people is definitely not recommended, nor running over a snake with a motor vehicle.

### Did you know?

*Snakes have tails too! Though a snake's body appears to have no tail, or perhaps to some appears to be all tail, officially, or according to the rules of science, a snake's tail begins at the anal opening. The length depends on the species and the size of the snake. Generally, the longer the snake, the longer the tail.*

### Did you know?

*Snakes do not have ears and cannot hear any sounds travelling through the air. Their 'warning sound' is the vibration of the ground created by footsteps or other movements, which they pick up by their inner ear. Cobras seen 'dancing' to music are probably coaxed into their actions by following the movement of the swaying instrument and the musician.*

**Did you know?**
*A snake's digestive system is highly dependent on the weather. Its digestive juices work properly only when the snake's temperature is high enough. Should the snake be in the process of digesting an animal, and the temperature outside dropped suddenly, the snake's temperature would also fall. The undigested food would eventually rot inside the snake's body, causing fermentation and death.*

## Red-bellied Black Snake *Pseudechis porhyriacus*

Native. Active by day. Venomous and potentially dangerous.

### Identification

Average length 1.5 metres, though may exceed two metres. Shiny black on top, with some brown markings around the snout. Lower sides are bright orange to red (rarely white), fading to a duller colour on the belly. The head is almost indistinct from the neck.

## Habitat

Leads a solitary life, though in favourable conditions large populations can be found. This excellent swimmer is usually found near water, such as near swamps, ponds, creeks, river-banks and lagoons, where food is in abundance. It is encountered far less frequently in dry forests.

## Breeding

Bears live young during summer or autumn. Litters vary from eight to 20, though can be as high as 40. At birth, young emerge from a membrane sac within minutes, though may take up to an hour. Young average 20 cm in length, and fend for themselves from birth.

## Food

Carnivorous. Frogs are the main diet. Also feeds on small mammals such as mice and rats, birds, lizards, and snakes, including its own species. Being a good swimmer, also catches fish and eels.

## Emergency food

Snakes do not eat frequently, and in an emergency can last without food for up to a few weeks, especially during cold weather. In captivity, feeds on whole mice and frogs.

## Hostile behaviour

It is generally not a hostile snake, and prefers to retreat when encountered. If it feels threatened or cornered, it may attempt to bluff its defence by hissing loudly, flattening the neck, and making bluffing strikes. At some level of irritation, has been known to emit an unpleasant-smelling discharge from the anal area.

## Handling and transport

Red-bellied Black Snakes are venomous. Handling, therefore, is not recommended, except by qualified people. Should it be necessary to capture the snake, see EXTRA HELP page 193. For further comments on snake handling and transport see SNAKES, HANDLING and TRANSPORT pages 143 & 145. For symptoms and treatment of snake bites see FIRST AID, SNAKE BITES page 12.

# Eastern Brown Snake *Pseudonaja textilis*

Native. Active by day. Venomous and dangerous.

*juvenile*

## Identification

Average length 1.5 metres, though may exceed two metres. A long, slender body, usually uniform in colour. Top colouring can be any shade of brown, from orange to black. Belly is yellow or orange, blotched with grey or a darker orange. The head is not distinct from the neck. Some juveniles may be quite distinct from the adults. They have a black head-band and may have 50 or more grey or black bands across their bodies. The bands disappear with age, though some adult snakes have been known to retain them, making them similar in appearance to a tiger snake.

### Habitat
Leads a solitary life. Found in a variety of habitats, including dry grasslands as well as wet forests, though shows a preference for drier areas. Common in rural areas, frequently entering barns in search of food.

### Breeding
Female deposits from ten to 30 eggs, in a secluded spot such as under a log, under rocks or in leaf mould. Young average 27 cm in length, and fend for themselves from hatching.

### Food
Carnivorous. Feeds on small mammals such as mice and rats, small birds, lizards, snakes, and frogs.

### Emergency food
Snakes do not eat frequently, and in an emergency can last without food for up to a few weeks, especially during cold weather. In captivity, feeds on whole mice and rats.

### Hostile behaviour
This snake is extremely aggressive and easily aggravated. It will raise its head, curving itself into an 'S' shape, and flattening its neck. It strikes repeatedly, with its mouth open. It can strike to a distance of its own body length.

### Handling and transport
Eastern Brown Snakes are venomous. Handling, therefore, is not recommended, except by qualified people. Should it be necessary to capture the snake, see EXTRA HELP page 193. For further comments on snake handling and transport see SNAKES, HANDLING and TRANSPORT pages 143 & 145. For symptoms and treatment of snake bites see FIRST AID, SNAKE BITES page 12.

## Dugite *Pseudonaja affinis*

Native. Active by day. Venomous and dangerous.

### Identification
Averages 1.5 metres in length, though may reach two metres. Top colour varies from grey or olive, to almost black, and is often spotted with black scales. Belly is a greyish-white, olive or yellow, and often spotted with darker flecks and a series of blotches. The head is small, indistinct from the neck, and often lighter than the body.

### Habitat
Leads a solitary life. Found in a variety of habitats, showing a preference for drier areas. Commonly found in urban areas, often entering houses and barns in search of food.

### Breeding
Female deposits from five to 25 eggs in loose soil or a hole in the ground. Young hatch in January, and measure about 18 cm in length. They fend for themselves from hatching.

### Food
Carnivorous. Feeds on small mammals such as rats and mice. Also eats small birds, lizards, and snakes.

J. C. Wombey/NPIAW

**Emergency food**
Snakes do not eat frequently, and in an emergency can last without food for up to a few weeks, especially during cool weather. In captivity, feeds on whole mice.

**Hostile behaviour**
When aggravated, raises its head, curving itself into an 'S' shape, and flattens its neck. Strikes repeatedly with its mouth open. It can strike to a distance of its own body length.

**Handling and transport**
Dugites are venomous. Handling, therefore, is not recommended, except by qualified people. Should it be necessary to capture the snake, see EXTRA HELP page 193. For further information on snake handling and transport, see SNAKES, HANDLING and TRANSPORT pages 143 & 145. For symptoms and treatment of snake bites see FIRST AID, SNAKE BITES page 12.

## Common Death Adder *Acanthopis antarcticus*

Native. Nocturnal. Venomous and dangerous.

### Identification
Average length is 40 to 60 cm, though may reach one metre. It has a short and thick body. Colour varies regionally. The top can be any shade of brown, from light to almost black, crossed with bands of a darker or lighter shade of body colour. Belly is cream to grey, with darker blotches. The broad head is very distinct from the neck.

### Habitat
Leads a solitary life. Prefers undisturbed bushland, where it often buries part of its body under leaves and debris. Its 'earthy' colour and burrowing habits make it hard to detect by passers by.

## Breeding

Bears live young during autumn. Litters vary from three to 20. Young average 13 cm in length, and fend for themselves from birth.

## Food

Carnivorous. Feeds on frogs, small mammals such as mice and rats, birds, and lizards. It lies quite still, often partially buried under debris, exposing only the tip of its tail. The tail tip looks like a worm, and when shaken vigorously, lures the prey to approach the snake. At that point, the snake strikes the victim.

## Emergency food

Snakes do not eat frequently, and in an emergency can last without food for up to a few weeks, especially during cold weather. In captivity, feeds on whole mice and rats.

## Hostile behaviour

Sits undetected, and makes no effort to move away with approaching footsteps, as other snakes do. This makes it easy to tread upon, and most bites are the result of that. It flattens its body, and strikes with an amazing speed.

## Handling and transport

Death Adders are venomous. Handling, therefore, is not recommended, except by qualified people. Should it be necessary to capture the snake, see EXTRA HELP page 193. For further comments on handling and transporting snakes see SNAKES, HANDLING and TRANSPORT pages 143 & 145. For symptoms and treatment of snake bites see FIRST AID, SNAKE BITES page 12.

# Eastern Tiger Snake *Notechis scutatus*

Other names: Mainland Tiger Snake

Native. Active by day, and in warm weather by night. Venomous.

**Identification**
Average length is approximately one metre, though may double that. Body is quite solid, and colouration varies regionally. The top can be any shade of brown, from light, through olive, to almost black. Most Tiger Snakes are crossed with about 45 yellowish bands, though unbanded specimens do occur. Belly is cream to yellow. The head is slightly distinct from the neck.

**Habitat**
Leads a solitary life, though in favourable conditions large populations of these snakes congregate. Covers a wide variety of habitat, from rainforests to dry land. Shows a preference for damp areas such as swamps, marshes and river banks.

**Breeding**
Bear live young during summer months. Litters average about 30, though can vary from 20 to 40. Young average 15 cm in length, and fend for themselves from birth.

**Food**
Carnivorous. Frogs are its main diet. Also feeds on fish, birds, lizards, and small mammals such as mice and rats.

**Emergency food**
Snakes do not eat frequently, and in an emergency can last without food for up to a few weeks, especially during colder months. In captivity, feeds on frogs and whole mice.

**Hostile behaviour**
Can become extremely aggressive if provoked. At that point, flattens its neck and body, and hisses loudly. Strikes low to the ground, so that most bites are to the foot or ankle.

**Handling and transport**
Tiger Snakes are venomous. Handling, therefore, is not recommended, except by qualified people. Should it be necessary to capture the snake, see EXTRA HELP page 193. For further information on handling and transporting snakes, see SNAKES, HANDLING and TRANSPORT pages 143 & 145. For symptoms and treatment of snake bite see FIRST AID, SNAKE BITES page page 12.

## Common Tree Snake *Dendrelaphis punctulatus*

Native. Brown: nocturnal; green: active by day. Mildly venomous (not life threatening).

### Identification
Average length 1 to 1.2 metres, though may reach 2 metres. Colour varies regionally. The top is uniform in colour, in any shade of green or brown. Belly is usually a shade of yellow, more intense in colour under the throat, though can be anything from white to green-yellow. The head is slightly distinct from the neck, and may be a darker shade than the body.

### Habitat
Leads a solitary life. Prefers timbered areas, spending most of its time in trees and shrubs. Often forages on the ground, frequenting rivers and streams. Shelters in tree hollows, rock crevices, and occasionally in suburban houses.

### Breeding
Female deposits from seven to 12 eggs in a secluded spot, such as under a log, under rocks or in leaf mould. Young average 33 cm in length, and fend for themselves from hatching.

### Food
Carnivorous. Frogs are its main diet. Also preys upon nesting birds, small lizards, and occasionally small mammals such as mice.

### Emergency food

Snakes do not eat frequently, and in an emergency can last without food for up to a few weeks, especially during cold weather. In captivity, feeds on whole mice and frogs.

### Hostile behaviour

Can bite repeatedly if provoked. It is mildly venomous. Usually resorts to a display of bluff behaviour, by raising its head and flattening its neck and body. At that point, the blue skin between its scales becomes visible. Has been known to emit an unpleasant-smelling discharge from the anal area when irritated.

### Handling and transport

Unless the snake has been positively identified as harmless, handling and transporting are not recommended, except by qualified people. Should it be necessary to capture the snake, see EXTRA HELP page 193. For further information on handling and transporting snakes, see SNAKES, HANDLING and TRANSPORT pages 143 & 145. For symptoms and treatment of snake bite see FIRST AID, SNAKE BITES page 13.

## Carpet Python *Morelia variegatus*

## and **Diamond Python** *Morelia spilotes*

Native. Mostly nocturnal. Non-venomous.

### Identification

Average length is two metres, though specimens up to four metres have been recorded. Diamond pythons are slightly shorter than carpet pythons. These pythons have two quite distinct colour forms.

CARPET: The top is any shade of brown, with blotches or broad bands in a lighter shade of body colour, and edged with black. Belly is cream or yellow, with darker blotches.

DIAMOND: The top is an olive-black, with clusters of yellow spots, forming diamond shapes, and edged with black. Belly is cream or yellow, with darker blotches. The head is quite distinct from the neck.

### Habitat
Leads a solitary life. Can be found in a variety of conditions, from forests where they shelter in trees or shrubs, to barren regions where they live in burrows made by other animals, such as rabbits. During the day, they rest in hollow logs, in tree hollows, and occasionally under house rafters or in barns.

### Breeding
Females deposit from 15 to 50 eggs in the summer, and coil their bodies around them to protect and incubate them. Young average 30 cm in length, and fend for themselves from hatching.

### Food
Carnivorous. Feed on small mammals such as mice, rats and rabbits, and occasionally lizards. Can accommodate large items of food.

### Emergency food
Snakes do not eat frequently, and in an emergency can last without food for up to a few weeks, especially during cold weather. In captivity, pythons feed on whole mice and rats.

### Hostile behaviour
Though non-venomous, they are quite unpredictable in temperament. Assume an 'S' position with the head and body, and can strike without warning.

### Handling and transport
Unless the snake has been positively identified as harmless, handling and transporting are not recommended, except by qualified people. Should it be necessary to capture the snake, see EXTRA HELP page 193. For further information on handling and transport of snakes, see SNAKES, HANDLING and TRANSPORT pages 143 & 145. For symptoms and treatment of snake bite see FIRST AID, SNAKE BITES page 13.

### Did you know?
*It is a misconception that pythons crush their victims to death. Having immobilised the victim, breathing is restricted by the python tightening its hold with the victim's every exhaled breath. Death comes by suffocation.*

### Did You Know?
*Though most people have an aversion to snakes, pythons have rendered themselves useful to man. They are encouraged to barns by some farmers, to keep down the population of rats and mice.*

# INSECTS

There are more species of insects than there are species of all the other animal groups put together. Although not always welcome, insects are an important part of the ecological cycle and contribute in many ways to nature's balance. Many keep down populations of other insects or creatures, many pollinate plants and some, like the honey bee, provide us directly with food. Even the termites, which are so feared by us for their destruction of wooden structures, are useful in their recycling of dead material in the forests.

## BEES

### How to deter bees and avoid being stung

There is no known reason why bees choose certain nesting sites in urban areas, such as in wall cavities, chimneys or in rockeries. There is no deterrent for this. If there are no flowers or flowering trees in your yard, you stand a smaller chance of being visited by foraging bees, though that is not a guarantee that bees will not nest on your property. Should you see bees hovering around one spot near a cavity on your property, particularly in early spring, they may be considering moving in. Spray the area with a domestic strength surface spray or hang a pest strip as close to the cavity as possible, to deter them from a lengthy stay. Ensure that air-vents in your house have a net covering to stop bees from entering to nest.

If you have a pool, bees may be attracted to it in summer, as they need water. Visiting bees inform other bees in the hive of any water sitings, and declare the area as their water supply. If the pool is occasionally covered, it won't be given the status of 'permanent water supply', as on occasions the bees will be forced to seek water elsewhere. An alternative water supply such as a bird bath or a fish pond, may also divert the bees' attention from your pool.

Outdoors, people may get stung by treading on a bee. Always wear shoes outdoors, particularly in areas rich in flowering clover plants, as clover is a favourite with bees. When on a picnic, a combination of bright clothing, strong perfume, perspiration odour and alcohol on the breath, is ill-advised when near bees. The mixture of bright colours with the smells may offend them, and make them hostile and more aggressive. Be careful about drinking out of a can outdoors, as a bee may have entered it. That can result in a serious sting to the throat. Use a straw or drink out of a glass.

### How to attract bees

The best way to encourage bees to your garden is to plant native flowers or flowering trees such as eucalypts, bottlebrushes, grevilleas and acacias to name a few. Clover is a favourite with bees. The bees in your garden will help pollinate the plants in it, which results in better quality flowers and vegetables. In summer bees need water; provision of a bird bath or a fish pond may help to deter bees from entering a swimming pool in the area to obtain water.

### Handling
Bees should not be handled with bare hands, except when they are in a state of low energy (when they cannot move). When in this state, they can be picked up by their wings, keeping clear of the abdomen where the sting is. If stung by a bee see FIRST AID, BEE STINGS page 14. Bees cannot be captured in mid-air and therefore must be trapped with a net or cornered onto a flat surface. Once on a flat surface, place a wide-necked jar over the bee. Slide a piece of paper or cardboard under the jar and the bee, trapping it inside. Roll the jar over to an upright position and secure a cover over it. The bee will keep in a closed jar for up to two hours, and in a jar with air-holes for up to a day. Lack of food, not air, is what eventually kills the bee. For prolonged keeping, place a drop of sugar and water into the jar. People with known allergies to bee stings should leave the capture of bees to others if possible.

### Preserving
Should it be necessary to capture a bee for positive identification, and the bee cannot be identified immediately, it may need preserving. Preserved specimens are preferable to dry ones. Capture the bee in a jar, and pour 75% ethyl alcohol or methylated spirits and 25% water over the bee, completely sumberging it. It should remain intact for several weeks. Dry specimens may be submitted only if the body is intact.

### Bees in the house
Apart from entering in search of food, the presence of bees in the house indicates that they may be considering settling within the property (for example in the wall or roof cavity, air vents, or chimney), or that they have already settled in the house. They are more likely to settle in spring or summer, as it is the breeding and swarming season. Bees which have accidentally ended up in a house will eventually die, unless released. They may take up to a day to die, and will most likely hover around a window, trying to leave through the glass.

Trapped bees should be released by opening a window or door, or left till their energy runs low (anything up to a day), and captured when they are less active. See BEES, HANDLING page 162. If there are a lot of bees in a house, the vacuum cleaner is handy for getting them out. Collect all the bees into the vacuum cleaner, and take outside where they can be released. If you know where the bees are entering, a domestic insect surface spray sprayed onto the area should stop them entering. It is a temporary solution, and the area may need spraying every day, or periodically.

It is wise to investigate why bees are entering the house. In such cases, there is a possibility that they are nesting within the house structure. If you suspect or identify a nest within a cavity, pest strips can be used to deter and eliminate the bees. The strip should be suspended on a piece of string, and lowered into the cavity/area where the bees are nesting. If a whole strip is too big, it can be cut into small pieces, and each piece lowered into the cavity separately. The strip should be put in place just before dark, when all the bees have returned to the nest and there is still some visibility. The strip should remain toxic for up to three months, and is effective only when placed into an enclosed area, whether it is used indoors or outdoors. If the wall cavity had bees nesting in it, the remains of wax are likely to attract another colony to settle there again. Therefore, having eliminated the bees, it is important to seal any entry points to the cavity or the same problem may occur.

There have been cases where bees nesting in a wall cavity have abandoned the nest without provocation. That may happen if the food supply in the area cannot sustain the hive population.

### A swarm of bees in the yard

A swarm is a natural way of dividing a crowded colony of bees into two colonies. The swarm is the half which has left the hive and settled temporarily, usually from 400 to 500 metres away from the original hive. A swarm may settle in your yard. If it is within your reach, to be safe, lock your pets away and advise everyone to keep away, though swarms are generally gentle in nature. While the swarm is on your property, a few worker bees from the swarm are out looking for suitable shelter for nesting. The swarm is therefore only transitory and will eventually leave your yard. Should it settle within your house structure, see BEES IN THE HOUSE page 162. Should you wish to have the swarm removed, see EXTRA HELP page 193 for people who may be able to help.

### Pet stung by bees or wasps

Generally, bee and wasp stings are not dangerous to pets. Dogs are more likely to get stung, as they tend to snap at insects. As in humans, some animals react more violently to being stung than others. If stung in the mouth or on the throat, breathing difficulty may develop as a result of the swelling, so the animal must be taken to a vet immediately. If stung on other parts of the body, let the animal settle before approaching. If stung by a bee, remove the sting with tweezers, and observe the animal. If there is a reaction such as swelling of the face or any part of the body, or frothing at the mouth, take the animal to a vet.

### Beekeeping

Beekeeping is not just for apiarists or people who live on farms. A hive can be set up and maintained in almost any backyard. If you keep bees, you will get satisfaction, honey, and extra pollination of your garden, resulting in better quality flowers and vegetables. Before proceeding, read and learn about bees and beekeeping. Contact a reputable beekeeper for advice and possible supply of a hive, bees, and necessary equipment. The beekeeper may be able to advise you of any amateur beekeeping societies or organisations in the area. It is wise to join one for advice on beekeeping and the possible use of some jointly-owned or hired equipment too expensive to buy for a hobby beekeeper.

Contact your local council to see if you should be guided by any special laws regarding setting up the hive, and once set up, the hive may need to be registered with your government agricultural body. See EXTRA HELP page 193.

Your beekeeper or members from the beekeeping society may be able to help in choosing the most suitable position for the hive,

within your yard. Choose the spot wisely, considering your neighbours and other structures, such as your clothesline, to fit in with the flight path the bees will take. Erect a water supply for the bees; it can be a bird bath or a fish pond. That will help deter them from frequenting your own, or the local, swimming pool to obtain water.
Note: One or a few hives will probably be run on a hobby basis. For it to become a profitable venture, further enquiries should be made.

**Did you know?**
*There are approximately 20 000 species of bees in the world, and about 3 000 in Australia. Most bees are solitary and do not live in colonies or hives. Bees make honey not for us but for other bees, though man has benefited greatly from this. Bees' usefulness does not end with honey. Their major contribution is pollination of plants, which results in better quality fruits and vegetables.*

## European Honey Bee *Apid millifera*

Introduced. Social. Active by day.
Dangerous only to allergic people.

## Identification

Approximate body length 12 to 15 mm, the queen being the largest and the workers the smallest. Body the bees reach adulthood. Before assuming foraging duties, worker bees first spend some time in the hive attending to babies and hive maintenance. In cold areas, brood production stops in autumn.

## Habitat

Being social, honey bees live in large colonies of approximately 5000 to 80 000 bees. In a natural state, colonies nest in hollows of trees, in caves, under hanging rocks, or anywhere where there is a crevice or shelter. In urban situations bees have set up nests in wall and roof cavities, air vents, chimneys, garden retainer walls or rockeries, under eaves, and in the controlled situation of beehives.
Hive: A hive or colony contains a queen bee for laying eggs; workers, sterile females which attend to the young, maintain the hive and collect nectar and pollen; and drones, which are fertile males for mating.
Swarm: A colony of bees with the same structure as in a hive, which are looking for shelter.

## Breeding

Spring to summer is the breeding season, peaking in October to November. Two types of eggs are laid by the queen: infertile, resulting in male drones, and fertile, resulting in female workers. There are more female eggs than male eggs produced. Each egg (larva), occupies a cell within the nest area, and is fed and attended to by worker bees. Within three weeks, the bees reach adulthood. Before assuming foraging duties, worker bees first spend some time in the hive attending to babies and hive maintenance. In cold areas, brood production stops in autumn.

## Food

Bees are nectar and pollen eaters. They rely on flowers for food. They gather nectar from blossoms with the aid of a long tongue, and collect pollen by brushing against the flowers with their hairy hind legs. Nectar and pollen can be gathered at the same time or independently. Human food such as jam, honey, split ripe fruit, and soft drinks, can also provide food for bees, though they are more likely to be attracted to the food during times of deficiency in natural food.

is thickly covered in fine hair, and hind legs have a pollen basket. Colour depends on breed. The most common colours are black or brown-black and yellow or orange-yellow markings on the abdomen. Native bees differ from the European honey bees by being much darker in colour on the abdomen and, in most species, being smaller. Most are a quarter of the size of the honey bee, and only a few are bigger.

# NATIVE WASPS

## Native Papernest Wasp *Polistes* spp

Native. Social. Active by day.
Dangerous only to allergic people.

### Identification

Approximate body length 11 to 20 mm. Body is slim and hairless, with a slim waist. Body colour is dark brown, with reddish or yellowish-orange bands across the abdomen. The antenna is short.

### Habitat

Lives in small colonies of approximately 10 to 20 wasps. Nests are found in relatively unconcealed spots, where they hang from branches of trees or shrubs, rock overhangs and, in urban situations, from walls, eaves of houses and off balconies.
Nest: nests are mushroom-shaped, averaging to 20 cm in diameter, and hang suspended by a single stem. Honeycomb cells, as in beehives, are visible from underneath the nest. The structure of the nest is paper-like material.

### Breeding

Breeds during the summer months. Fertilised female wasps emerge out of winter hibernation during early spring, and lay one egg per cell in the nest. The pupae are attended to by adult wasps. They emerge as adult males and fertile females.

### Food

Ominivorous. Feeds on insects and their larvae, caterpillars (some of which are pests to crop growers), spiders, nectar, and fruit juices.

### Handling

The only reason for handling these wasps should be if their capture is necessary for positive identification.

If their nest can be seen, then that is sufficient identifier, as it is unique to these wasps. Where the nest cannot be seen and the wasps are active, a wasp must be captured for positive identification. Wasps should never be handled with bare hands as they can sting repeatedly, unlike bees which can sting only once. If stung by a wasp, see FIRST AID, WASP STINGS page 15. They cannot be captured in mid-air, and must be cornered onto a flat surface or sprayed with a domestic insecticide and collected dead. If on a flat surface, place a wide-necked jar over the wasp. Slide a piece of paper or cardboard under the jar and the wasp, trapping it inside. Roll the jar over to an upright position and secure a cover, or cover with a lid. The wasp will keep in a closed jar for up to two hours, and a jar with air-holes for up to a day. A wasp sprayed with insecticide for collection should not be presumed dead; it should be scooped into a jar with an object, not by hand. People with known allergies to bees and wasps should leave the capture to someone else, if possible.

**Preserving**
See BEES, PRESERVING page 162.

**Native papernest wasps nesting on your property**
If the nest is on your property but far away from human contact, the wasps should cause you no problems and should not put anyone in danger. Consider leaving the nest alone, and observing nature at work. Should the nest be too close to a window or overhead in your yard, it could become a problem as the wasps can become aggravated by human presence too close to the nest. The nest could be eliminated, and it is best done early evening or after dark, when the wasps have retired for the night. The safest method is to use domestic strength insecticide spray. Lay some newspaper sheets under the nest. Spray a burst of insect spray directly into the nest cells, and leave. The wasps should drop dead onto the paper. The following morning, knock the nest off its stem and let fall onto the paper. Wrap the dead wasps and the nest in the paper, and dispose of it in a bin.

Another method of elimination is to use a plastic bag, though you may risk being stung. Again in the evening or after dark, place a plastic bag over the nest, enclosing the nest and the wasps in it. Break the nest stem and let the nest fall into the bag. Seal the bag securely, and dispose of in a manner which won't risk releasing the wasps, such as the bag bursting in the garbage.

**Pet stung by a wasp**
See page 163.

**Did you know?**
*As the name suggests, these wasps construct their nests from paper — in fact they mastered the art of paper-making long before man did. The wasps collect tiny fragments from dried stems of plants, and mix them with saliva into a pulp (cellulose).*

# INTRODUCED WASPS

## European Wasps *Vespula germanica*

P. Hadlington

Introduced. Social. Active by day. Potentially dangerous.

### Identification

Average body length of the workers 12 to 15 mm. The males and the queen are larger, and may reach 20 mm. Body is stocky and hairless. Body colour is bright lemon-yellow, with distinctive markings on the abdomen, black V-shapes with a dot on either side. The antennae are relatively long, being about half body size.

Usually, all the wasps within a colony, except for the fertilised queen, die off during the cold winter months. Australian warm winters have at times caused exceptions to the rule, and colonies have survived the winter.
Nests: usually built in concealed areas such as in tree trunks, in the ground and, in urban environment, in rockeries, in retainer walls, wall cavities, sub-floor areas, under eaves, and in buildings within a yard.

Nests are concealed within crevices, usually showing only an external entrance hole opening. They are greyish in colour with a papery appearance, and are built to mould into the crevice. Less

Flick

common, external nests are usually oval-shaped, and in extreme cases can be several metres high. The opening of some nests is covered in a grey papery material, making it difficult to detect.

### Breeding

Spring and summer is the breeding season. Fertilised queen wasps emerge from winter hibernation and lay eggs. In three to four days the eggs hatch and are attended to by the queen. Several weeks later, the wasps emerge from pupal stage as adults, and take over the work of the nest.

### Food

Omnivorous. Feed on a variety of foodstuffs. In urban situations, scavenge on barbecues, leftover pet food, ripe fruit, garbage in bins, sweets and sweet drinks. Also feed on nectar, grubs, and dead animal flesh.

### Handling

The advice on handling is don't, unless absolutely necessary. Only experienced people should attempt to approach a nest closely or destroy one. If it is necessary to capture a wasp for positive identification, it is safer to concentrate on a single wasp that's hovering instead of one within a group. Do not attempt to capture a wasp near the nest, as it may be more aggressive and your chances of being stung are greater. Do not handle with bare hands, as wasps can sting repeatedly, unlike bees which can sting only once. If stung by a wasp, see FIRST AID, WASP STINGS page 15. As a wasp cannot be captured in mid-air, it must be against a flat surface or cornered, so it can be killed prior to capture. Spray the wasp with a domestic insect spray. Scoop into a jar with an object, not by hand. Cover the jar securely or put the lid on. If you have a known allergy to bees and wasps, leave the capturing to someone else.

### Preserving

See BEES, PRESERVING page 162.

### How to deter European wasps and avoid being stung

Unfortunately there is no deterrent for these pest wasps. Lack of flowers or flowering trees in the yard is no guarantee against their visits, as nectar is not the only food they consume. Being omnivorous, the smell of meat and sweets at barbecues and picnics may attract them, so caution should be taken. When outdoors, it is safer to drink out of a glass or through a straw, instead of out of a can. A wasp may enter the can while you're not looking, which can result in a serious sting to the throat. Always wear shoes when outdoors, as you risk stepping on a wasp. Be careful when eating ripe fruit, in case a wasp has settled on it.

Around the yard, remove all uneaten pet food and keep all garbage bins closed with tight-fitting lids. If you have fruit trees, any fallen fruit lying on the ground may attract wasps. Remove the fruit as soon as possible. Be careful when handling the fruit, as a wasp may have settled on it. Screening all windows and doors helps keep wasps away from household foods, which may be displayed or cooking near an open window. As these wasps raid hives and kill bees, be particularly aware of their activity if you have a hive on your property.

### European wasps nesting on your property

If you suspect a nest of European wasps on your property or in any spot where they are endangering you, you must organise to have it destroyed. It may involve catching one wasp for identification. See EUROPEAN WASPS, HANDLING page 169. If all the wasps are too close to the nest, it is safer to call someone to come out to identify the wasps and the nest, instead of trying to catch a wasp. Keep a distance of at least two metres from the nest, especially during warmer months when the wasps are more active and easily aggravated, and never touch the nest. Nests can be destroyed only by people experienced in that field, such as beekeepers or pest control companies.

### Comparison of honey bees to European wasps

The following may help distinguish honey bees from European wasps.
Body colour: Bees are black-brown and orange-yellow. European wasps are black and lemon-yellow.
Legs: Bees have black legs. European wasps have yellow legs.

*EUROPEAN WASP*

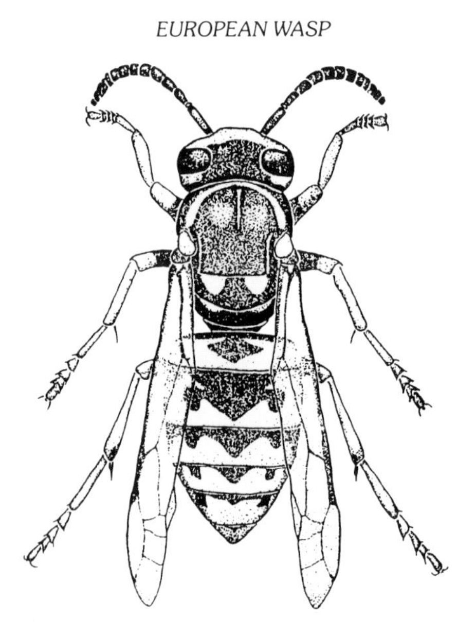

*HONEY BEE*

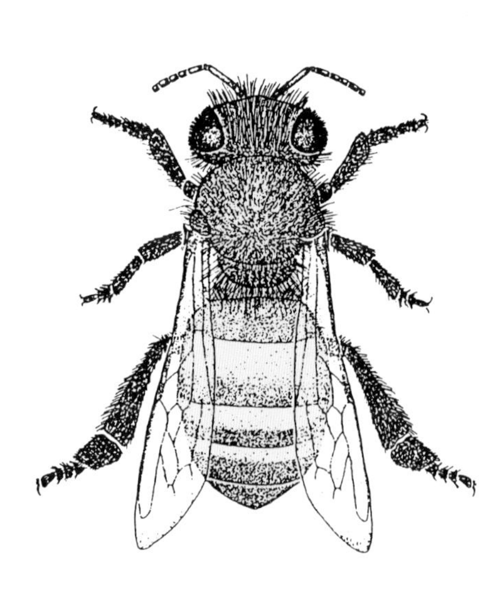

Antennae: Bees have short antennae. European wasps have long antennae, of half their body length.
Abdomen: Bees have plain bands of orange-yellow on a black-brown abdomen. European wasps have black V-shaped markings with a dot on each side, on a yellow abdomen.
Hind legs: Bees have pollen baskets on hind legs. European wasps have slim legs with no baskets.

### Comparison of native papernest wasps to European wasps

The following may help distinguish native papernest wasps from European wasps.

Body colour: Papernest wasps are orange and brown. European wasps are black and bright yellow.
Body shape: Papernest wasps are slender, with a narrow waist. European wasps are more stocky.
Antennae: Papernest wasps have short antennae. European wasps have long antennae, of half the body length.
Wings: Papernest wasps have coloured wings. European wasps have clear wings.

### Pet stung by a wasp

See page 163.

**Did you know?**

*Since their accidental introduction in 1959, these wasps have been a source of concern for their multiple stings, their aggressive nature, and their habit of robbing honey from beehives. They have also spread a hatred of all wasps throughout the community. Fear and lack of knowledge cast a suspicion on all yellow, orange or black flying insects. This has resulted in the destruction of many useful, harmless, native wasps. The Native Papernest Wasp, which has the misfortune to bear a resemblance to the European Wasp, has taken the biggest beating. Get with it! Learn about your wasps!*

# TERMITES

## Termites Coptotermes *Coptotermes scinaciformis*

Other names: White Ants

Native. Social. Subterranean. Destructive.

### Identification

Workers: 3 to 4 mm long, wingless with an unpigmented creamy body with a large head. Blind and sterile.
Soldiers: to 6 mm long, wingless, unpigmented creamy body. Brownish head with prominent jaws. Blind and sterile.
Reproductives: called alates. Body length without wings to 8 mm, with the wings to 13 mm. Body is more pigmented than the worker's, in a

*a bad infestation*

brown colour. Narrow wings are cast off at end of breeding season. Fertile. Queen: as the body fills with eggs, it can swell up to 50 mm in length.

**Habitat and life cycle**
Live in colonies of up to a few million termites. A colony is formed from a pair of reproductives, see BREEDING CYCLE page 173. The reproductive pair chooses a nest site. It must be in a humid and dark environment near a source of food, most common places being in tree stumps, in the ground or in a domestic situation under houses. (Some species of termites nest in trees or build mounds on the ground). As the nest population increases in size, new food supplies must be reached to sustain the

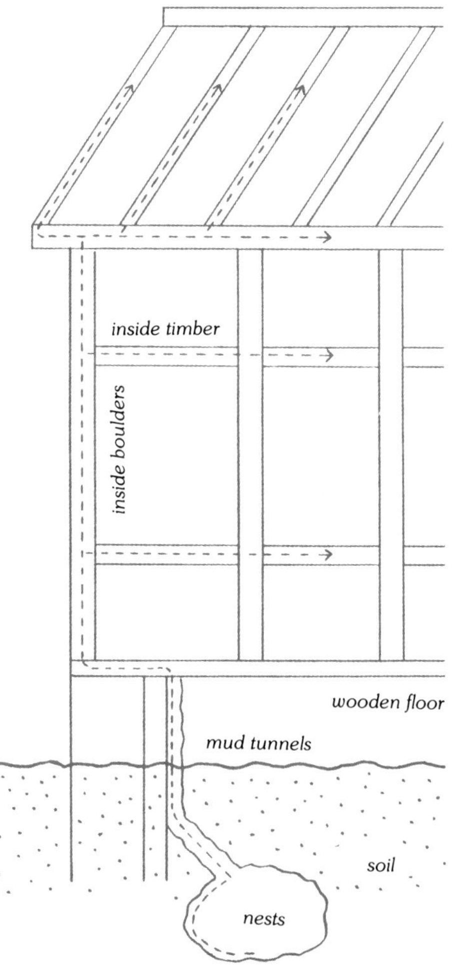

colony. Tunnels are excavated underground linking the nest with a food supply, which can be a dead tree, timber fence or timber posts within a house structure. Having reached the timber, it is eventually completely hollowed out, preserving the outer structure intact to maintain the humidity and shelter from light and predators.

If the food supply cannot be reached through the tunnels underground, mud tunnels are built along structures which cannot be hollowed out, such as along brick piers or on concrete. The tunnels are constructed from a mixture of faeces, soil particles and saliva, and can be from one to several centimetres wide. They act as a link between the nest and the source of food. As the colony grows in size, new food supplies must be found constantly, creating many tunnels in the process. A house attacked by termites can be up to 50 metres away from the nest.

Nest: earthern structure, with a central nursery about the size of a football. The first thing to be established in a new colony.

Tunnels: underground excavations, joining the nest with a food supply.

Mud tunnels: above-ground constructions, joining the nest with a food supply.

## Breeding cycle

Breeding takes place during the warmer, more humid months of late spring or early summer, usually in November to December. Reproductive termites (alates) swarm out of the nest and scatter. Casualties are high, and the remaining few pair off, fall to the ground, drop their wings, and search for a suitable nesting site. Having mated, the queen lays the eggs from which workers, soldiers and potential reproductives (in case the mating pair dies) emerge. The young are attended to by the queen. When mature, the workers attend to the young, forage for food, feed the queen and the soldiers, and excavate the tunnels. The soldiers guard the colony. Once the colony is established, about five years after the mating of the original pair, eggs are laid and new reproductives emerge. Those are the future kings and queens of new colonies. The reproductives leave the nest upon reaching maturity.

## Food

Basically vegetarian. Feed on any materials containing cellulose, such as wood and timber. Some timber is more readily attacked than others and a preference is shown for timber attacked by fungi, due to the available source of protein. Apart from the fungi, protein is obtained by eating the bodies of dead termites. Only non-living parts of trees are attacked, such as dead branches and the heartwood of trees. Paper, cotton, carpets and leather have also been attacked by termites.

## Handling/preserving

Termites are incapable of inflicting a wound, and therefore can be handled safely. Their small size makes handling of individuals difficult, and tweezers can be used to pick out individual termites. Should capture be necessary for positive identification, it is the soldiers (ones with the dark heads), or the winged reproductives that should be captured, as they are more distinctly marked. Place at least a dozen in a jar, and pour 75% methylated spirits or ethyl alcohol and 25% water over the termites to preserve. Dry specimens are not as useful for identification. Wash your hands after handling the termites.

### How to deter termites

There are certain steps which may be undertaken to discourage termites from finding a property suitable for nesting.

Chemical barriers: Prior to pouring of concrete slabs or after the foundations have been completed for suspended floors, the soil can be treated with a chemical barrier by a reliable pest control company. If the slab has been poured over untreated soil, it can be extremely expensive and inconvenient to protect it from termites, as they can enter the house through tiny cracks and expansion joints. Fences can also be protected, by treating the soil around the fence posts; however, on joint properties, don't forget to liaise with your neighbour.

Termite shields or caps placed over foundation piers and walls will not stop termites crossing to gain access to the wood; however, they may help deter them and assist in early detection of termite mud tunnels. Do not store any timber or paper cartons under or against houses, as they create a favourable environment and may encourage termites. Remove all left over timber from under houses after construction. Also, treat or remove all tree stumps found during construction. Provide adequate ventilation and good drainage under suspended floors and around buildings, as they can create humidity and moisture which help encourage termites.

Although no timber is safe from termites, some timbers are more resistant than others; for example, cypress pine and river red gum are not as readily attacked as untreated pine or oregon. Treated pine is termite-proof and suitable for decking, though consider the durability of the joists and piers. The decking may be solid, but the piers may be vulnerable to attack.

Regular inspection is most important. Inspect the sub-floor area, the roof cavity, and timber fixtures within the house at least once a year. Concrete slabs should be inspected around the edges and ends of walls within the building. Should you suspect a termite attack, disturb the colony as little as possible, as it may affect the possible treatment.

### How to detect termite activity in a dwelling

All subterranean termite activity, no matter how high within the house structure, must stay in contact with the ground for moisture. (Drywood termites are one exception, as they do not need ground contact for moisture.) Termite activity can be traced to the original nest, which is usually in

Flick

the ground either beneath, or up to 50 metres outside, the house. The most obvious sign of termite activity is mud tunnels, constructed to link the nest with a new source of food. They can be from one to several centimetres thick, and run up brick piers, woodwork, along pipes, walls, or over ant cappings. They are grey to brown in colour, and if broken will show termite activity by exposing the termites. The termites will seal the hole within a few hours. Should no termites appear and the hole remain unsealed after 24 hours, the nest may be inactive and the mud tunnels should be scraped off.

Termites in timber which is in direct contact with the ground are much harder to detect, and sometimes go unnoticed till a lot of damage has been done. Very often the first discovery of infested timber is made when the pile is disturbed for some reason. The inside of such timbers is quite hollow due to the excavated tunnels, and if poked with a screwdriver, the structure would crumble easily. A screwdriver can be used to test for termite activity in timber within a house. If pressed gently into the timber, attacked timber would sink the screwdriver easily or make a hollow sound if knocked upon.

If a termite tunnel is broken or termites are exposed by breaking into the timber, a faint clicking noise can be heard coming from inside. It is the soldier termites making a warning sound. When inspecting concrete slabs for termite activity, check all settling in cracks and expansion holes, and look for mud tunnels around the edges of the slab.

### Flying reproductive termites

In the warm, humid months of late spring and summer, the reproductive termites leave their nests and disperse. A small percentage of them survive and succeed in forming new colonies. It is a natural occurrence, and their presence near your house should not be cause for alarm. Should they be flying into the house and their presence concerns you, a spray with a household insecticide spray should eliminate them. Once inside a house, there is little chance of them causing a problem. A more likely place for them to settle would be under the house. If they were successful in setting up a colony, it would take approximately five years from the time they landed before they could be detected. If concerned, the only protection against reproductives settling on a property is not to create an environment favourable to them. See HOW TO DETER TERMITES page 174.

### Termites in the yard

There may be termite activity in the ground, in a tree or a mound, inside or close to your yard. Not all species of termites attack houses, and termites in a yard are not always cause for alarm. Most people cannot distinguish various species of termites, and the matter should be investigated by experts. You must obtain a few termites to present for identification, or call an expert to the site. As all termite colonies are sealed, you must break the surface to collect a few termites. The soldiers (ones with the darker heads) or the flying reproductives are the ones most suitable for positive identification.

See TERMITES, HANDLING page 173. A small puncture with a screwdriver is all you will need to break the surface of a tunnel, after which the soldier termites should surface near the hole. Having collected about a dozen, leave the hole, as the workers will seal it within a few hours. Having identified the species, you may learn that it is harmless, or you may need to call out professionals to deal with the problem.

**Did you know?**

*Though often called white ants, termites are not ants. Their closest relatives are cockroaches. Termites are able to destroy wood more quickly than any other insects, though not all species of termites attack houses. (In Australia there are approximately 300 species of termites.) Their contribution to nature is to recycle dead material in the forests.*

# COCKROACHES

## German Cockroaches *Blattella germanica*

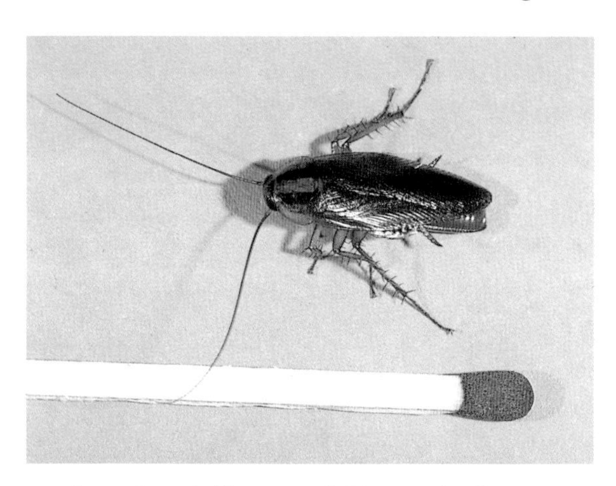

Introduced. Nocturnal. Domestic. Pest.

**Identification**

Adult body length 10 to 15 mm. Body is oval and flat, for quick retreat into narrow cracks and crevices. Overall body colour is light brown. Wings lie flat on the back, and this species rarely flies. Legs are well developed for running, and are covered with protective spikes. The antennae are long and curve along both sides of the body.

Nymphs: smaller than adults and darker in colour. They lack wings and are often mistaken for other insects.

## Habitat

Live in small groups, or infestations of thousands of cockroaches. Mostly found in urban areas in interior of buildings. Warmth, moisture and food are their requirements, and those are best met in food-handling areas such as kitchens and pantries. During the day they hide in concealed crevices around the kitchen sink, hot water system, behind appliances such as stoves, refrigerators and dishwashers, and in and under cupboards and drawers. (The larger American cockroaches do not need elevated temperatures and live in sub-floors, roof voids and outdoors in gardens and gutters.)

## Breeding

Female lays 30 to 40 eggs in a pillow-shaped egg capsule. For 16 to 30 days the capsule is carried attached to her body. It is deposited in a dark crevice just prior to hatching. The nymphs hatch in an incomplete state and become adults within 6-12 weeks, after several moultings.

## Food

Omnivorous and a scavenger. Feeds on any material of plant or animal origin, such as household leftovers and crumbs, meat, dairy and starchy products, sweets, labels on bottles and tins, fabric, paper or cardboard, leather, hair, glue, book bindings, woollen fabrics, and wallpaper including the glue. Can survive without water for up to a month.

## Handling

As they are good runners and extremely fast, catching live cockroaches is next to impossible. To capture for positive identification, the cockroach may need to be killed first. If cornered, spray an insecticide spray directly onto the cockroach. If none available, the cockroach may have to be squashed with a broom or a flat object, but this may dismantle the body making identification more difficult. As cockroaches are potential carriers of disease organisms, handling with bare hands is not recommended. Pick up with a tissue or brush with a dustpan and broom. Wash your hands after handling.

## Carriers of disease

Cockroaches are known to carry disease organisms such as those that cause food poisoning (salmonella), gastroenteritis, tuberculosis, dysentery, hepatitis, and many more. Some people are allergic to cockroaches. Contact with cockroach faeces or food contaminated with faeces may bring about a skin allergy or an asthma attack in asthmatics. In such cases, medical aid should be sought.

## How to deter cockroaches

German cockroaches thrive in areas which provide food, warmth, moisture, and hiding places. To deter them, food and hiding places must be eliminated or kept to a minimum. This can be achieved by keeping food-preparing areas scrupulously clean and maintaining the house. Wipe up all food spills and remove crumbs and leftover food immediately, before they settle into cracks or build up, such as grease around ovens and hotplates. Store opened food in the refrigerator or in containers with tight-fitting lids. Avoid leaving dirty food utensils and leftovers on plates overnight. Remove all uneaten pet food. Keep garbage bins tightly closed and clean them regularly. Reduce shelter areas by filling any holes and cracks around the house, such as loose tiles, loose skirting boards, and gaps around kitchen cupboards. Fit screens on all windows and draught-strips on all doors. Check for cockroaches or eggs within any cartons brought into the home.

### Cockroaches in the house

The presence of cockroaches in a house signifies that the premises have met some or all of their requirements for survival: food, warmth, moisture, and shelter. Being nocturnal, their activity and the size of infestation is best observed after dark. Should the cockroaches be active in daylight hours, it may be a sign of a large infestation where overcrowding has occurred. Another sign of a large infestation is an unpleasant odour caused by a buildup of secretions from the mouth and cuticle. The first measure is to apply extreme cleanliness in food-handling areas. See HOW TO DETER COCKROACHES, page 177.

The next step is to eliminate the cockroaches. A variety of products are available on the market for the purpose, such as dusts, sprays, baits, strips, and surface sprays. Read the labels and follow the instructions carefully. Do not treat food-handling areas such as breadboards and bench tops with toxic materials, unless the label instructs you otherwise. Apply the product in any potential hiding places such as around sinks, behind appliances such as stoves, refrigerators and dishwashers, around water pipes especially hot ones, in cupboards (empty first), around skirting boards and in any cracks and crevices. Eliminate any future hiding places by sealing off as many cracks and crevices as possible. Should the infestation be large, or your attempts to deal with the problem unsuccessful, contact a reputable pest control company to deal with the problem.

**Did you know?**

*Of all the creatures in the world, the cockroach is the most adaptable and the most successful survivor. It has withstood the test of time, dating back some 300 million years, when according to fossil records it was very much the same as it is now. Based on that, we can assume that it will probably still be here long after man has gone.*

# SPIDERS

Many people are fearful of spiders, yet most are timid and their venom is generally not seriously harmful to humans. Australia's Redback and Funnelweb Spiders are two exceptions, so caution should be exercised if you are unsure what species you are dealing with.

Spiders are major predators of insects. Their methods of catching them are not limited to trapping in webs: they may also chase or jump upon their prey. Many inject venom into their victim, too.

In Australia today scientists are experimenting with 'milking' certain native species so that their venom can be used to make biological insecticides that may protect crops, such as cotton, from insect infestation.

In order to grow, spiders must cast off their outer skins as reptiles do. This 'moulting' procedure takes place several times in a spider's lifetime. When young spiders go through their moult to become full-grown adults they emerge as either male or female. Most spiders cease moulting at that stage but some females, including the funnelwebs, continue to moult. Moulting takes several hours, during which the spider becomes quite helpless and in danger of drying out. The old cuticle splits in various places and the spider pulls itself out, leaving the discarded cuticle in one piece. The new skin is soft and colourless, becoming harder and gaining colour after a few hours.

## How to identify a spider

Though the most reliable way to identify a spider is from a specimen, there may be a time when help is needed over the phone. The following points are a guide when trying to identify a spider. Where was it found? In the house, yard, bush, in the ground, on a tree, around windows, or elsewhere? Type of web (if any): was it circular, funnel-shaped, a tangled mass, or other? Size of the spider: saying large, medium or small, is not enough; compare the spider to a coin, such as 5¢, 10¢, 20¢, 50¢, or compare it to other objects, such as a pea.
Explain whether the size includes the legs or just the head, shoulders and the abdomen.

Colour: is the spider one even colour or many colours? Is the body shiny or dull? Are there any distinctive marks? If the spider is dead, turn over and check the abdomen. Is the abdomen color different to the dorsal surface? Are there any distinctive markings on the abdomen? Texture: is the spider hairy, smooth or a combination of the two? Legs: Are they thin, sticky, long or short? Spider's movement: Did it jump, run, or otherwise?

There may be other questions you may be asked to answer.

## Preserving spiders

If the spider cannot be identified immediately upon catching, it may need to be preserved. Dried specimens are not as useful for positive identification. To preserve, pour 75% ethyl alcohol or methylated spirits diluted with 25% water into a jar, completely submerging the spider. The spider should keep for some weeks.

## A spider in the pool

The spider in your pool may be a funnelweb or a trapdoor spider. The males of both species roam at night, during the breeding season of summer and autumn, looking for a mate. Since they are hard to distinguish, it is safer to assume that you are dealing with a potentially dangerous spider and apply caution, as for a funnelweb spider. These spiders fall into pools either because they are attracted to moisture (only funnelwebs), or because they fall in by accident. They fall to the bottom, and the air trapped around their bodies may give them a glittering appearance or the appearance of being in a bubble of air.

Once in the pool, the spiders are not capable of climbing out, and a prolonged stay means eventual death. There are no accurate figures as to how long they can stay underwater alive, though up to 24 hours is a guideline suggested by some spider experts. The spider should still be treated with caution, as data from a major pest control company revealed a spider showing movement after four days in the pool. For safety reasons, never presume a spider dead, no matter how long it has been in the pool.

Fish the spider out with the pool scoop, and let it fall into a wide-mouthed jar if you wish to have it positively identified. See EXTRA HELP page 193, for a list of people who may help identify the spider. If you wish to squash the spider, scoop it out and place on a hard surface, with the scoop net still covering it. Squash with a heavy object such as a rock, over the net.

Some spiders may end up in the filter basket. It is not a good idea to empty the basket with bare hands, without checking it for spiders first.

Note: if immediate identification of the spider is not possible, it should be preserved in the jar in 75% ethyl alcohol or methylated spirits, and 25% water. Dried specimens are not as useful for positive identification.

## Pet bitten by a spider

Though funnelweb and redback spiders can be dangerous to man, nature has been kinder to domestic animals by providing them with a natural immunity against such bites. In most cases, you will never know that your pet has been bitten, as it won't show any symptoms. Should you witness such a bite to your pet, no first aid procedures are needed, and a trip to the vet is not necessary, unless any unusual symptoms appear.

## Holes in your yard

A small hole in the ground in your yard does not always mean the presence of a Funnelweb Spider. If there is no evidence of silk lining around or inside the hole, then it could belong to a cicada. If it is silk-lined but the entrance to the burrow lacks the threads of silk stretching out to join the surrounding area, then it could belong to a Trapdoor Spider or another spider. Position of the hole is also an indicator. Funnelwebs like damp, more sheltered conditions; trapdoors prefer open, dry conditions. Not all burrows are occupied. If you are concerned about the occupant, you could flush it out by pouring boiling water or petrol down the hole. A word of caution, you may be hurting a harmless and useful spider, or be confronted by a dangerous aggravated spider. By flushing the occupant out, you probably won't kill it, just get a better look for identification. At that point, you may wish to kill it, capture it to have it positively identified, or leave it alone and call out professional pest exterminators. To capture, see HANDLING on pages relevant to the species.

# GROUND-DWELLING SPIDERS

## Sydney Funnelweb Spider *Atrax rubustus*

Native. Nocturnal. Venomous.

### Identification

Approximate body length, head, shoulders, abdomen: 15 to 20 mm male, 20 to 30 mm female. Body is an even colour of glossy black, though can be a dark or reddish brown. Underside has reddish hair. Males differ from females by having a smaller abdomen, longer legs and spur on each of the second front legs. Spinnerets (finger-like projections at the rear of the abdomen from which the spider spins its thread) can be clearly seen on both sexes.

### Habitat

A ground dweller, found in burrows constructed in favourable conditions of sheltered, damp, low temperature, humid spots. Found under logs and stones, in rock crevices, at bases of trees, in leaf litter, or in man-made situations such as stored timber or building materials. Burrow: a silk-lined tube, up to 30 cm deep, with one or two entrance holes. Top of burrow can

be distinguished by trip-lines: strands of silk stretching out to join the surrounding rocks and debris. Purpose of these is to alert the spider of passing prey. They are characteristic of funnelweb spider burrows.

**Breeding**
Summer to autumn is the breeding season. Female lays 80 to 200 eggs in a suspended pillow-shaped egg-sac, which she guards. After hatching, the young remain in the burrow for several months. In late summer or autumn they disperse, during damp weather, when the ground is soft for digging burrows. Only a few spiders live to reach adulthood.

**Food**
Insectivorous and carnivorous. Main diet includes a variety of invertebrates such as moths, beetles, crickets, millipedes, slaters, snails, ants, cockroaches, and other spiders. Small lizards and frogs are also eaten. 'Trip-lines' at the entrance to the burrow are used to alert the spider of the presence of prey, whereupon the spider seizes it, dealing the fatal bite. The victim is then carried into the burrow for storage and consumption.

**Hostile behaviour**
You know when a funnelweb spider has biting on its mind, by the stance. The front half of the body is raised off the ground, with the two front legs high up in the air. It will then lean back on the other legs, and lunge forward to strike the victim in a forward, downward strike.

**Handling**
Funnelwebs are venomous spiders. Handling, therefore, is not recommended, except by qualified people. Should it be necessary to capture it for positive identification, under no circumstances should it be done with bare hands. For symptoms and treatment of funnelweb bites, see FIRST AID, FUNNELWEB SPIDER BITES page 15
Never assume that the spider is dead, even if it appears so. Wait for it to move, or coax it onto a flat surface. Place a widemouthed jar over the top of it, trapping it underneath. So much the better if the jar can be handled with long

tongs, allowing greater distance between you and the spider. Slide a piece of cardboard under the jar and the spider. Fold the sides of the cardboard around the jar neck or, holding the cardboard firmly in place, roll the jar to an upright position. Keep your hands away from the jar neck at all times. Secure the jar cover, or cover with a lid, taking care not to release the spider. Once at the bottom, it cannot escape, as it cannot climb up the glass. Coaxing the spider into a jar using a stick is another method of capture; however, it is best left to experienced people or less dangerous spiders. Under cool conditions, the spider will stay alive in the jar for at least a day.

### How to deter funnelweb spiders and avoid being bitten

Knowledge about this spider plays an important role in its deterrance.

Educate your family, particularly the children, about this spider. Show them pictures, talk about first aid procedures, tell them to leave all spiders alone. Find out if your area has a high funnelweb incidence. Develop safety-conscious habits: wear gloves and shoes when gardening, do not walk barefoot at night. Keep your yard tidy and your shrubs trim. Rockeries make excellent burrow sites, and stored or piled junk, piled leaves or timber are also possible sites for burrows. If storage of materials is necessary, store on an elevated platform, about half a metre off the ground.

Screen all windows and doors; ensure the screens are tight-fitting to stop spiders entering during their night wanderings. Fit a draught-strip to doors. Funnelwebs can climb vertical walls made of brick, cement or timber; they cannot climb glass surfaces. Check your pool for spiders before diving in, see SPIDER IN THE POOL page 180. Shake out clothing and shoes which have been left outside, particularly if they were there overnight, or were left on the ground.

Digging, excavating, and landscaping disrupts the spiders' habitat, causing them to become restless and wander. Avoid such activities if possible during the funnelweb breeding months of summer and early autumn. Excessive rain may have a similar effect on the spiders. Avoid widespread application of insecticides, as if the spiders have not been directly 'wetted' by the insecticide, it may cause them to wander. Treating potential harbourage areas directly is more effective. Be aware that every summer and autumn, male funnelwebs wander at night looking for a female to mate with. Make those your priority times for spider awareness.

### A funnelweb spider in the house

Although it is rarely that a funnelweb spider will wander into your house, there are several reasons why it might. Its habitat may have been disturbed, or its burrow may have been flooded out due to heavy rain. Someone may have brought it in on their clothing, which was left outside overnight. During the breeding season of summer and autumn, male funnelwebs wander at night in search of a mate, and may accidentally wander into a dwelling. Do not panic, reserve your energy for some plan of action. Without losing sight of the spider, arm yourself with a long object such as a broomstick. Try to sweep the spider out the door, or onto an area where it can be cornered and squashed. Do not step on the spider; squash it with a long object such as a broom. If you are not alone, call out for someone to come with a broom, while you keep track of the spider's location.

Note: contrary to popular belief, these spiders cannot jump.

**A funnelweb spider in the yard**

Depending on the severity of the problem, you may need to call out professional pest exterminators to deal with spider problems, or you may be confronted by a single funnelweb spider in your yard, which you may decide to eliminate yourself.

If the spider resembles a funnelweb but has not been positively identified as one, it is safer to assume that you are dealing with a potentially dangerous spider and treat it as a funnelweb. If the spider is on a flat surface, a hard blow with a long-handled object, such as a spade, should result in instant death. Do not attempt to squash it with your foot, as you place yourself in a potentially dangerous situation. You can also throw a heavy object such as a rock onto the spider, from a distance. Be careful when lifting the rock to check if the spider is dead. Keep your distance, and lever the rock with a stick to lift.

If the spider is in the funnel, spray household insecticide into the hole and plug it securely with dirt or surrounding debris. Pack into the hole, and push down with a spade, broom handle or a stick, to collapse the hole. You may choose to dig up the hole and the spider. Wear shoes, gloves and leg covering. You may have to dig deep as some burrows are up to 30 cm deep. Be prepared to kill the spider as soon as you have reached it. Again, the spade can be used.

Never poke spider holes with fingers or sticks, as you may aggravate the spider and risk being bitten.

**Did you know?**

*Funnelweb spiders cannot jump. Their quick jerky movements and tendency to rear up prior to biting may have contributed to the belief that they can. Funnelwebs are Australia's, and one of the world's, most dangerous spiders.*

*They are not just ground-dwelling. Some species live in trees. There are approximately 35 species of Funnelwebs throughout the eastern coast of Australia. Most of them are dangerous to man.*

### Comparison of funnelweb spiders to trapdoor spiders

Though at first glance these spiders appear to look similar, there are distinguishing features which can be used to differentiate them.
Colour: funnelwebs are more likely to be a shiny black or a shiny brown. Trapdoors are more likely to be a dull brown.
Spinnerets: funnelwebs have long spinnerets. Trapdoors have short ones.
Spur: male funnelwebs have a spur on the second set of legs. Male trapdoors have a spur on the first set of legs.
Habitat: funnelwebs are more likely to be found in sheltered areas. Trapdoors are more likely to be found in open areas.

## Brown Trapdoor Spider *Misgolas rapax*

Raymond Mascord

### Identification

Approximate body length, head, shoulders, abdomen: 12 to 20 mm male, 20 to 30 mm female. Body is covered in fine hair, and is usually a dark brown, though can be almost black. The abdomen is dull. Top of abdomen can be banded with yellowish stripes, though they may be indistinct. Males differ from females by having a smaller body, longer legs, and by having large palps resembling boxing gloves. Spinnerets are short on both sexes.

Native. Nocturnal. Mildly venomous.

*found on the East Coast*

## Habitat

A ground-dweller. Lives in burrows, and is often mistaken for funnelweb spiders. The burrows of trapdoors are built in more open, sunny, drier conditions, and on level ground or in banks, not undercover.
Burrow: a silk-lined tube, up to

Raymond Mascord

40 cm deep, sometimes protruding up to 3 cm above ground. Brown Trapdoor Spiders, unlike most others of the species, do not build a door to the burrow. The rim of the burrow lacks the strands of silk characteristic of the burrows of funnelweb spiders.

## Breeding

Summer to autumn is the breeding season. Female lays 35 to 50 eggs in a suspended pillow-shaped egg-sac, which she guards. After hatching, the young remain in the burrow for several months. In late summer or autumn they disperse during damp weather, when the ground is soft for digging burrows. Only a few spiders reach adulthood.

## Food

Basically insectivorous. Main diet includes a variety of invertebrates, such as moths, beetles, crickets, millipedes, slaters, snails, ants, cockroaches, and other spiders. All prey is caught near the burrow, and taken into the burrow for consumption.

## Hostile behaviour

Trapdoor spiders seem to copy the funnelwebs not just in looks, but also in the stance prior to biting, though they are not as aggressive. Front half of the body is raised off the ground, with the two front legs high up in the air. The spider then leans back on the back legs, and lunges forward to strike the victim in a forward, downward motion.

## Handling

Brown Trapdoor Spiders are mildly venomous and should be handled like funnelwebs, see SYDNEY FUNNELWEB SPIDER, HANDLING page 182.

## How to deter trapdoor spiders and avoid being bitten

Due to the similarity of this spider to the dangerous funnelweb spider, when encountering the trapdoor one can never be sure that it is not a funnelweb. For that reason, it is safer to assume that you are dealing with a potentially dangerous spider, and apply procedures for funnelweb, see page 183.

**Did you know?**

*As the name suggests, most trapdoor spiders build doors to their burrows, though some other spiders do as well. The door serves many purposes. It is a handy thing to close just as the enemy is approaching; it makes a wonderful ambush surprise for catching prey, not forgetting the air-conditioning factor. Contrary to the name, the Brown Trapdoor Spider does not build a door to its burrow. Of all the trapdoor species found in Australia, it is the only one that doesn't.*

## Redback Spider *Latrodectus hasseltii*

Native. Nocturnal. Venomous.

Dr Mike Gray

### Identification

Approximate body length, head, shoulders, abdomen: to 5 mm male, to 10 mm female. Female body colour is a shiny black, most often with a chracteristic red band along the dorsal abdomen, though the band can be pink or orange. Males differ from females by being smaller, usually more brown in colour, having no red band, and having white markings on the abdomen. Both sexes have long, thin, shiny legs. The identifying feature of this spider is an orange or red hour-glass shaped marking on or under the abdomen.

## Habitat

This spider prefers to build its web in populated areas. Any site will do, provided it is sheltered, dry, quiet, and offers some sort of food supply. Some suitable spots are in rubbish or litter, in old tins, under houses, under rocks, around window and door frames, on pot plants, under guttering eaves, around piled or stored timber or building materials, in old machinery, under broad-leafed plants such as vegetables, behind car bumper bars, and an old favourite, around outdoor toilets.

Web: untidy looking, loosely spun tangle of silk, with strands of silk stretching out to join the surrounding area. Though hidden, at the top of the web is a tube-like retreat where the spider rests and keeps eggs.

## Breeding

Summer to autumn is the breeding season. Female lays 50 to 300 eggs in spherical-shaped egg-sacs, which she guards. They hatch in ideal weather conditions of warmth and dampness, and disperse if conditions continue to be favourable. Should they remain in the web due to unfavourable weather conditions, they become cannibalistic, gradually eliminating each other.

## Food

Insectivorous and carnivorous. Main diet includes a variety of insects such as beetles, crickets, ants, other spiders, and flies. Small lizards and frogs are also eaten. When prey becomes entangled in the web, the spider seizes it, dealing the fateful blow. It is then carried to the retreat for storage and consumption.

## Handling

Redback Spiders are venomous. Handling, therefore, is not recommended, except by qualified people. Should it be necessary to capture the spider for positive identification, under no circumstances should it be done with bare hands. For symptoms and treatment of redback bites, see FIRST AID, REDBACK SPIDER BITES page 16. Never presume that the spider is dead, even if it appears so.

As most webs back onto a surface or are in a crevice, the spider may have to be taken out with some of the web. Stick a twig inside the funnel of the web, and twist till the web and the spider wind around it. Carefully scrape off the twig into a jar, or preferably place twig and spider in the jar.

If the spider is against a flat surface, place a wide-mouthed jar over the top of it. Having trapped it under the jar, slide a piece of cardboard or firm paper under the jar and the spider. Fold the sides of the paper around the jar neck or, holding the cardboard firmly in place, roll the jar to an upright position. Secure a cover or cover with a lid, taking care not to release the spider. Once at the bottom, it cannot climb up the glass.

Coaxing the spider into a jar using a stick is another method of capture; however, it takes more skill and the risk of being bitten is greater. Under cool conditions, the spider will stay alive in the jar for at least a day.

## How to deter redback spiders and avoid being bitten

These spiders have adapted particularly well to man's environment, and prefer and seem to thrive in disturbed areas. They are also hardy and hard to control.

Educate your family, particularly the children, about these spiders. Show them pictures, talk about first aid procedures, tell them to leave all spiders alone. Keep your yard tidy and free of stored material, empty tins, corrugated iron, piled timber or piled building materials, as they are ideal nesting places. If storing material is necessary, keep to a minimum and check periodically for webs. Also check barbecue areas and garden sheds. Remove any webs as soon as they are found.

As most webs back onto a surface or are in a crevice, the best method of elimination is to squash the web, with the spider in it, with a stick or a block of wood. Or, stick a twig inside the funnel in the web, and twist till the web and the spider wind around it. Then, step on the twig to squash the spider. The webs can also be sprayed with a domestic insecticide spray. Spray directly onto the spider or into the funnel in the web. Wear gloves and shoes when gardening, collecting timber or clearing out garden sheds.

Wear protective clothing when climbing under houses, as the dry sheltered area under houses makes a suitable habitat for redbacks. Unsewered toilets are a favourite with Redback Spiders. Keep them clean and check regularly for spiders. Ask a reliable pest control company if they can be of assistance.

**Did you know?**

*Redback Spiders are more common around human habitats than in the bush. They are closely related to the dreaded American Black Widow Spider.*

## Black House Spider

*Badumna insignes* (formerly *Ixeuticus robustus*)

Raymond Mascord

*found on the East Coast*

Other names: Window Spider

Native. Nocturnal. Mildly venomous.

**Identification**
Approximate body length, head, shoulders, abdomen: 10 to 15 mm male, 15 to 20 mm female. Body is covered in fine hairs, mostly a shade of black, dark brown, or grey. The abdomen may be marked in a lighter shade than the body colour, such as grey or cream.

**Habitat**
Likes to build its web in sheltered areas such as crevices in trees, or amongst rocks. Commonly found around houses, with webs around windows and doors, under guttering, in cracks of walls, on trellises, and in fences.

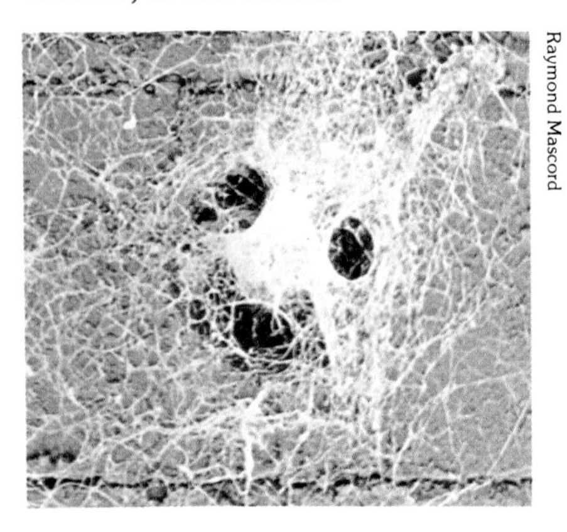

Raymond Mascord

Web: untidy-looking mass of fine silk, with one or more funnel-like entrance holes leading to a central retreat. As the spider 'maintains' the web, the web increases in size and can reach up to 30 cm across.

**Breeding**
Eggs are stored in an egg-sac, built on the side of the retreat. The spiders take about two years to mature.

**Food**
Basically insectivorous. Main diet includes a variety of invertebrates, such as moths, ants, beetles, flies, and any other insects unfortunate enough to become entangled in the web, where they are caught.

**Handling**
Though this spider is only mildly venomous, it is capable of inflicting a painful bite and should be treated with caution. Should it be necessary to capture it for positive identification, it should not be done with bare hands. For details of handling see REDBACK SPIDER, HANDLING page 188. For symptoms and treatment of spider bites see FIRST AID, BLACK HOUSE SPIDER BITES page 16.

**Black House Spiders on your property**
The main disadvantage of having this spider on your property is its rather untidy web, which is not exactly a work of art and tends to accumulate dust. Fortunately, the spider is more useful than the web. It consumes flies and other annoying insects which can be more of a problem than the spider. Though capable of giving a nasty bite, it is not interested in attacking you, and does not usually enter a house. Most of the webs are built in sheltered places, away from human activity, see HABITAT. Should you still find its presence too much to bear, the web and the spider can be easily eliminated. If on a flat surface, brush off or squash with a broom. If the web is in a crevice, it can be squashed with a twig or a stick, or a twig can be poked inside the funnel in the web and twisted till the web and the spider wind around it. Then, step on the twig to squash the spider inside. The spider and the web can also be sprayed with a domestic insecticide spray. Spray onto the spider or into the funnel in the web.

## Huntsman Spider *Isopeda immanis*

Other names: Triantelope, Tarantula
Native. Nocturnal. Mildly venomous.

*found on the East Coast*

### Identification
Approximate body length, head, shoulders, abdomen: male to 30 mm, female to 40 mm. Can reach up to 10 cm across legs. Body colour is a combination of greys and browns, with a dark brown stripe along the centre of the abdomen. Leg colour is a combination of grey and brown spots and stripes. Overall body appearance is 'flattened'. Body is covered in fine hair, and the first set of legs are longer than the hind set.

### Habitat
These wandering spiders do not build a web or a permanent retreat. They usually shelter under loose bark of trees or under rocks or logs. During rainy weather, they enter houses where they are frequently seen moving sideways along walls or ceilings. Indoors, they shelter behind household objects such as curtains, pictures, and blinds. Outside, they shelter in crevices of fences and under eaves.

### Breeding
Summer and autumn is the breeding season. Female lays 120 to 150 eggs in an oval, suspended egg-sac, which she guards till the young emerge. The young stay with the female for some time, before dispersing.

### Food
Basically insectivorous. Main diet includes a variety of large insects such as moths, beetles, crickets,

cockroaches, and a variety of bark-dwelling beetles and other insects. Prey is caught by the spider pouncing on it, and dealing the fatal bite.

## Handling

Though this spider is not considered dangerous, it is a swift mover and will bite if provoked or roughly handled. It should not be handled with bare hands. If bitten, see FIRST AID, HUNTSMAN SPIDER BITES page 16. Should its capture be necessary for positive identification or to relocate it, it should be coaxed onto a flat surface if not already on one. Place a wide-mouthed jar over the spider, trapping it underneath. Slide a piece of firm paper or cardboard under the jar and the spider. Holding the paper/cardboard firmly in place, roll the jar over to an upright position. Keep the jar covered as the spider can climb up the glass and out. Another method of capture is to coax it into the jar with a twig or a stick. The spider does not always co-operate and this method is not easy. Under cool conditions, the spider will stay alive in the jar for at least a day.

## How to deter Huntsman Spiders

Its large, hairy body and unusual sideways movements have many people afraid of this relatively harmless spider. Unfortunately, it has a habbit of popping up inside houses, particularly after rain. There is no way to deter these spiders from entering the house, though tight-fitting screens over windows and tight-fitting doors with a draught-strip may help. If there are no trees in your yard, or there are only a few trees, and those are away from your house, chances are reduced of you having Huntsman Spiders on your property. Take comfort in the fact that this spider is harmless, and it can be beneficial. It won't attack you or mark your walls, and in return it will help keep down the population of annoying insects. Consider catching the spider (see HUNTSMAN SPIDERS, HANDLING page 192), and relocating it outside, if its presence inside bothers you. If compassion fails you, the only solution is the broom stick or an insecticide spray.

### Did you know?

*When one thinks of spiders, one thinks of webs. However, not all spiders rely on webs for food or shelter. The Huntsman Spider is one of those. Its extremely flat body is well adapted to life in the narrow space between a tree trunk and the bark. The spider's two front legs are longer than the two at the rear, for extra manouvrability. That allows the spider to move sideways as well as forward, which is a handy feature when it comes to speedy exits from confined places.*

# EXTRA HELP

Use these lists as a guide. Further research may be needed, and policies may vary in different States. Voluntary organisations and free services are under no obligation, and help depends on availability of time, and skilled staff on the premises at the time. It is advisable to record the phone numbers of useful organisations, as it may save time in an emergency. Most voluntary organisations would gladly accept monetary donations, though these are optional; indeed, some organisations may be badly in need of funds.

## GENERAL

**State Fauna Authority (phone)**

.........................

Government authority on wildlife and flora. Every State Authority has a head office and branch offices. Will not handle wildlife, but will answer queries on wildlife, supply pamphlets on some species, or direct to other points of help. Can issue a licence (free) to temporarily hold protected native animals. Can advise on which species it is legal to hold in captivity. Is interested in reports of malicious deeds to wildlife.
**State head offices:** New South Wales National Parks and Wildlife Service; ACT Parks and Conservation Service; Department of Conservation, Forests & Lands, Victoria; Queensland National Parks and Wildlife Service; South Australian National Parks and Wildlife Service; Western Australian Department of Conservation and Land Management; Conservation Commission of the Northern Territory; Department of Lands, Parks and Wildlife (Tasmania).

**Veterinary surgeons (phone)**

.........................

Private enterprises which will, as a rule, treat native animals. Members of the Australian Veterinary Association (AVA) may treat native fauna free of charge, as per AVA policy, while other vets may vary individually. Some vets have more experience with wildlife than others; some surgeries are better equipped than others. Always ring first to check if the vet has the expertise, time, and facilities, and whether he or she will charge for the service. Some vets sell wildlife feeding equipment and advise on rearing of native animals.

**Museum (phone)**

.........................

Will not handle animals, but will answer general queries (not on rearing) about wildlife or direct to other points of help. May be able to identify mammals by their droppings. Pamphlets are available on some species. Will accept bodies of dead birds, for research or display. Can identify most insects from a specimen. May have contacts for herpetologists.

**Zoo (phone)**

....................

An authority on wildlife. Will answer general queries about wildlife or direct to other points of help. Some offer an enquiry service on rearing and care of native wildlife. Others answer queries, depending on availability of time and experienced staff. Pamphlets on some species and problems are available in some zoological gardens. May have contacts for herpetologists.

**RSPCA (phone)**

....................

Animal-oriented organisation, with a head office and branch offices in every State. Authorised to rescue, treat and release native animals. Can direct to other animal-oriented organisations, which can help with wildlife. Interested in reports of any malicious deeds to wildlife.

**Amateur/voluntary animal organisations (phone)**

....................

There are many organisations dealing with wildlife. Each one provides a different service, and is guided by individual sets of rules and policies. Check your phone book, looking for words such as Amateur, Animal, Australian, Birds, Club, Field, Herpetologists, Mammal, Marsupial, Native, Orphaned, Ornithologists, Reptiles, Society, Wildlife, to name but a few. There are also individual experts in animal care.

**Wildlife rescue (phone)**

....................

At the time of publication, Sydney had two voluntary wildlife rescue organisations. WIRES (Wildlife Information and Rescue Organisation Service) and AWARE (Australian Wildlife Ambulance Rescue Emergencies), the southern group. They rescue and rear native animals, and answer queries on rearing. For rescues of larger reptiles, they may put you in touch with a herpetologist. They may supply special food and utensils formulated for animal rearing. State Fauna Authorities may have names or phone numbers of similar organisations in other States.

**Animal parks (phone)**

....................

Animal parks vary in size and capabilities. Some may have facilities to accept injured or orphaned fauna, while others can answer queries on animal rearing and injuries. All problems are dealt with depending on the availability of time and trained staff on the premises at the time of enquiry. Seasonal overloads such as peak tourist season or maintenance of the premises are not good times to pursue enquiries. Some animal parks are equipped to take sick, orphaned or injured koalas. Some may have phone numbers of herpetologists.

**Society for Growing Australian Plants (phone)**

....................

A group of flora enthusiasts, who work in harmony with wildlife. There are State and regional offices, and local study groups. They can advise on the best plants for a particular area, and ones most suitable for attracting wildlife. They can also offer pamphlets on the subject.

**Local councils (phone)**

.....................

Service provided by local councils varies. Most should be familiar with the fauna and flora in their area. Some may have a list of local beekeepers and beekeeping organisations. Consult the council before erecting a beehive, in case special regulations apply. Some may have the addresses of herpetologists in their areas.

**Pet shops (phone)**

.....................

These private enterprises vary as to goods sold and expertise of the staff. Apart from pet food, some of which may be suitable as emergency food for birds and native animals, pet shops may sell utensils for feeding young birds and animals. Some may offer advice on rearing birds and animals, and on where to obtain possum traps.

**Local police (phone)**

.....................

When a native animal is endangering the public (such as a magpie that is constantly on the attack), or when an animal is suffering and cannot easily be put down (such as an injured kangaroo on the road), see if your local police can help. In some cases, permission from the State Fauna Authority may be required before dealing with the animal. Some local police stations may have contacts for herpetologists.

**Electricity supply authority (phone)**

.....................

The body governing the supply of your electricity will probably have a fleet of cherry-pickers for access to electrical wires. Although under no obligation, and depending on availability of staff and equipment, they may be of assistance to animals caught in electricity wires. If the supply of power is interrupted as a result, inform the authority immediately.

**Pest control companies (phone)**

.....................

Pest control companies are private enterprises and charge for their services. They can help with some mammal, bird and insect problems. They are especially helpful with possums, rodents, introduced birds, termites, spiders, some bees and wasps and in some cases, bats. They may also be able to identify mammals by their droppings or insects from a specimen. Some major pest control companies sell or hire noise machines to deter birds on a large scale, and for termites they may provide soil treatment prior to or during construction.

**Chemists (phone)**

.....................

Chemists sell baby foods, baby vitamin drops, and glucose powder, which may be useful for feeding mammals or birds in an emergency, or when rearing. Items such as eye droppers and syringes are also sold by chemists, and are useful tools for feeding young mammals. Enquire about pet teats. Chemists also sell iodine and mercurochrome drops, which can be useful for treating minor wounds on reptiles and tortoises.

# MAMMALS

**Koala Preservation Society of NSW Inc. (phone)**

..................

A group of volunteers experienced in treating and rearing koalas. They have set up the Koala Hospital and Study Centre for the purpose. Though based at Port Macquarie, they are happy to advise over the phone for the most suitable places to accept injured or orphaned koalas in areas outside Port Macquarie. They can also advise on emergency procedures, and generally how to help koalas.

**Equipment Hire (phone)**

..................

These private enterprises vary in size and services provided. Some hire possum traps.

# BIRDS

**Australian Pigeon Fanciers Protection Association (phone)**

A group of pigeon enthusiasts, with a head office in NSW, and many regional clubs. They do not deal with feral pigeons, only racing or show pigeons, but will answer queries on any pigeons. They unite banded lost pigeons with their owners, and should be notified of malicious deeds to pigeons. They may also advise on pigeon rearing and problems caused by pigeons.

**Bird Breeders (phone)**

..................

Some bird breeders may be able to advise on bird rearing or treating injured birds. Most should supply food for birds, mealworms, and feeding utensils.

# REPTILES

**Australian Herpetological Society (phone)**

..................

A voluntary organisation of reptile enthusiasts. Will advise about reptile care, reptile problems, and will perform reptile rescues where necessary. Societies are found throughout Australia. In Sydney, at the time of publication, society policy was to not be listed in the phone book. The following bodies may have listings of the society's members: some local councils, some police stations, museums, some animal parks, animal rescue organisations, zoos.

# INSECTS & SPIDERS

**State Agriculture Department (phone)**

..................

A government department. In NSW, it is the NSW Department of Agriculture and Fisheries. Their main services concern insects, such as bees, wasps and cockroaches, and to a lesser degree, spiders. If they are unable to help, they will direct to other points of help, and

can provide pamphlets on some species of insects. They have a list of beekeepers and beekeeping organisations. They can identify some insects from specimens.
In other States they are: ACT/SA/NT/WA/Tasmania: the Department of Agriculture. In Victoria and Queensland, they are the Department of Primary Industry. Their services may differ from those provided in NSW.

**The State Forestry Department (phone)**

....................

In NSW and the ACT this is a separate government department, known as the Forestry Commission of NSW. It is an authority on any wood-related insects, such as termites. Pamphlets are available on some species, and there is a charge for some. Termites can be submitted for positive identification, and general queries on termites will be answered, or the enquirer directed to other points of help.
In some States, the department is combined with the NSW equivalent of National Parks and Wildlife. Victoria: Department of Conservation, Forests & Lands, Victoria. Queensland: the Forestry Department. South Australia: Woods and Forests Department. Western Australia: Department of Conservation and Land Management. Northern Territory: Conservation Commission of NT. Tasmania: the Forestry Commission. Services may vary throughout the States.

**Beekeepers (apiarists) (phone)**

....................

Services may vary. Most beekeepers should be able to advise on beekeeping, identify a bee or wasp from a specimen, and advise on bee and wasp problems. Most should be able to relocate a swarm, and attend to problems of bees or wasps nesting within a house structure or yard. There may be a charge for house calls.

**Amateur beekeeping organisations (phone)**

....................

Bee enthusiasts have formed various clubs and organisations, which can be tracked through beekeepers, local councils, or in the phone book. Look for Apiarist, Bee/beekeeping, Hive, Honey, or under the name of your region/city/State. Services offered vary. Most should advise on beekeeping and problems with bees. Some may have time-sharing of some equipment necessary for maintaining a hive.

# BIBLIOGRAPHY


Adams, George *Birdscaping Your Garden* (1987) Rigby, Sydney
Archer, M., Flannery, Dr T. *The Kangaroo* (1985) Weldon Pty Ltd., Sydney
Burton, Robert *Bird Behaviour* (1985) Granada Publishing, London
Cayley, Neville W. *What Bird Is That?* (1987) Angus & Robertson, Sydney
Cayley, Neville W. *What Mammal Is That?* (1987) Angus & Robertson, Sydney
Clyne, Densey *Wildlife In The Suburbs* (1986) Oxford University Press, Melbourne
Cogger, Harold G. *Australian Reptiles in Colour* Reed Books, Sydney
Cogger, Harold G. *Reptiles and Amphibians of Australia* (1983) Reed Books, Sydney
Covacevich, J., Davie, P., Pearn, J., (eds) *Toxic Plants and Animals* (1987) Queensland Museum, Brisbane
Davey, Keith *Australian Lizards* (1977) Periwinkle Press, Sydney
Davies, Wally (ed.) *Wildlife of the Brisbane Area* (1984) The Jacaranda Press, Brisbane
Goode, John *Insects of Australia* (1987) Angus & Robertson, Sydney
Gow, Graeme F. *Snakes of Australia* (1983) Angus & Robertson, Sydney
Griffiths, Ken *Reptiles of the Sydney Region* (1987) A Three Sisters Production Pty Ltd., Winmalee, N.S.W.
Hadlington, Phillip W., and Johnston, Judith A. *An Introduction to Australian Insects* (1982) New South Wales University Press, Sydney
Hadlington, P. and Gerozisis, J. *Urban Pest Control* (1985) New South Wales University Press, Sydney
Healy, Anthony and Smithers, Courtenay *Australian Insects in Colour* (1983) A H & A W Reed Pty Ltd, (Sydney)
Kinghorn, J.R. *The Snakes of Australia* (1967) Angus & Robertson, Sydney
Macdonald, J.D. *Understanding Australian Birds* (1982) A H & A W Reed Pty Ltd, Sydney
Macdonald, J.D. *Birds of Australia* (1984) Reed Books, Sydney
Macwhirter, Pat (ed.) *Everybird* (1987) Inkata Press, Melbourne
Mascord, Raymond *Australian Spiders in Colour* (1983) A H & A W Reed Pty Ltd, Sydney
Moffat, Averil *Handbook of Australian Animals* (1985) Bay Books, Sydney
Morcombe, Irene and Michael *Australian Mammals in Colour* (1979) Reed, Sydney
Officer, Brigadier Hugh R. *Australian Honeyeaters* The Bird Observers Club, Melbourne
*Reader's Digest Complete Book of Australian Birds* (1983) Reader's Digest Services Pty Ltd., Sydney
Russell, Rupert *Spotlight on Possums* (1980) University of Queensland Press, Brisbane
Sindel, Stan *Australian Lorikeets* (1987) Singil Press, Sydney
St John Ambulance Australia *Australian First Aid Manual* Ruskin Press, Melbourne
Strahan, Ronald (ed.) *The Australian Museum Complete Book of Australian Mammals* (1983) Angus & Robertson, Sydney
Sutherland, Dr Struan K. *Venomous Creatures of Australia* (1985) Oxford University Press, Melbourne
Swanson, Stephen *Lizards of Australia* (1976) Angus & Robertson, Sydney
Watts, C.H.S. and Aslin, H.J. *The Rodents of Australia* (1967) Angus & Robertson, Sydney
Weigel, John *Care of Australian Reptiles in Captivity* (1988) Reptile Keepers Association, Gosford, N.S.W.
Wheeler, Jack *The Care of Sick and Orphaned Native Birds and Animals* Geelong Field Naturalists Club

PRINTED NOTES:

AGFACTS Department of Agriculture NSW
Flick Pest Control
Forestry Commission of New South Wales
National Parks and Wildlife Service NSW
Queensland National Parks and Wildlife
The Australian Museum
The Society for Growing Australian Plants

# INDEX